Parallax of Growth

For Kathrine

Parallax of Growth

The Philosophy of Ecology and Economy

Ole Bjerg

polity

Copyright © Ole Bjerg 2016

The right of Ole Bjerg to be identified as Author of this Work has been asserted in accordance with the UK Copyright, Designs and Patents Act 1988.

First published in 2016 by Polity Press

Polity Press
65 Bridge Street
Cambridge CB2 1UR, UK

Polity Press
350 Main Street
Malden, MA 02148, USA

All rights reserved. Except for the quotation of short passages for the purpose of criticism and review, no part of this publication may be reproduced, stored in a retrieval system, or transmitted, in any form or by any means, electronic, mechanical, photocopying, recording or otherwise, without the prior permission of the publisher.

ISBN-13: 978-1-5095-0623-1
ISBN-13: 978-1-5095-0624-8(pb)

A catalogue record for this book is available from the British Library.

Library of Congress Cataloging-in-Publication Data

Bjerg, Ole, 1974- author.
Parallax of growth: the philosophy of ecology and economy / Ole Bjerg.
Cambridge; Malden, MA: Polity Press, [2016] | Includes bibliographical references and index.
LCCN 2015025357| ISBN 9781509506231 (hardback) | ISBN 9781509506248 (pbk.)
LCSH: Ecology–Philosophy. | Environmental economics.
LCC QH540.5 .B54 2016 | DDC 577.01–dc23 LC record available at http://lccn.loc.gov/2015025357

Typeset in 10.5 on 12pt Sabon
by Toppan Best-set Premedia Limited
Printed and bound in the United Kingdom by Clays Ltd, St Ives PLC

The publisher has used its best endeavours to ensure that the URLs for external websites referred to in this book are correct and active at the time of going to press. However, the publisher has no responsibility for the websites and can make no guarantee that a site will remain live or that the content is or will remain appropriate.

Every effort has been made to trace all copyright holders, but if any have been inadvertently overlooked the publisher will be pleased to include any necessary credits in any subsequent reprint or edition.

For further information on Polity, visit our website: politybooks.com

Contents

Introduction: Lenin at the Supermarket — 1

PART I THE LOGY OF ECO — 11
1 Balance of Nature — 13
2 Ecology Beyond Biology — 28

PART II THE NOMY OF ECO — 53
3 How *is* the Economy? — 55
4 The Market Theory of Value — 85
5 The Fantasy of Growth without Bounds — 109

PART III 'ECONOMY OR ECOLOGY? YES, PLEASE!' — 141
6 The Need to Grow — 143
7 The Desire to Grow — 162
8 The Drive for Growth — 192

Conclusion: 'It's the Money, Stupid!' — 227

Notes — 234
Bibliography — 241
Index — 247

Introduction:
Lenin at the Supermarket

The answer to the dual crisis of ecology and economy is found on a commercial billboard at a Stockholm underground station. An ad for Willy:s supermarket (Figure 0.1) shows the picture of a bag of green-labelled groceries with the following text written on the bag: 'Given the choice between ecology and economy we do not think that you should have to choose. Willy:s. Our business plan: The cheapest lunch box in Sweden.'

Of course, the immediate reading of the ad merely suggests that here is yet another option for the capitalist consumer, who wishes to save the planet from global warming and other environmental dangers without having to fundamentally change her habits of consumption. At Willy:s, the consumer can buy environmentally friendly products at a cheap price. This is green capitalism at its best.

The obvious point of the ad from Willy:s is of course that it is possible to have both ecology and economy at the same time. You don't have to choose. However, on closer philosophical inspection, the ad offers the possibility of yet another reading. This reading may serve as a first indication of the way that this book intends to approach the growth crisis of contemporary capitalism. Slavoj Žižek may provide us with the tools to open this alternative reading of the ad in the form of an old joke from the Soviet Union. The background of the joke is that Lenin would always encourage young people to educate themselves with the slogan 'learn, learn, learn'.

> Marx, Engels, and Lenin are asked whether they would prefer to have a wife or a mistress. As expected, Marx, rather conservative in private

Figure 0.1 Willy:s billboard ad

matters, answers, 'A wife!' while Engels, more of a bonvivant, opts for a mistress. To everyone's surprise, Lenin says, 'I'd like to have both!' Why? Is there a hidden stripe of decadent jouisseur behind his austere revolutionary image? No, he explains: 'So that I can tell my wife that I am going to my mistress, and my mistress that I have to be with my wife…' 'And then, what do you do?' 'I go to a solitary place to learn, learn, and learn!'[1]

The joke shares the same initial structure as the ad since both open with a choice between two seemingly mutually exclusive options: ecology or economy, wife or mistress. The immediate point of the Willy:s ad is that it provides a third option, which allows the consumer to choose both positions. Buying cheap organic bananas, the consumer may choose ecology and economy at the same time. In comparison, the joke opens not three but four different positions:

(1) I prefer a wife. This is the position adopted by Marx; (2) I prefer a mistress. This is the position adopted by Engels. Now the third possibility branches off into two positions: (3) I'd like to have both wife and mistress because this gives me access to the decadent *jouissance* of sleeping with two women. This is 'everyone's' immediate interpretation of Lenin's choice. Yet Lenin provides the final position: (4) I'd like to have both so that I can tell either one that I'm with the other, which allows me to be alone and learn, learn and learn. This fourth 'Leninist' option is the logical route that we are going to follow in this book in the effort to study the relation between ecology and economy within the context of contemporary capitalism. Transposing the logic of this fourth option into the domain of ecology and economy, we arrive at the theoretical approach that I shall be referring to as eco-analysis.

The dilemma between ecology and economy is not merely a choice that faces the contemporary consumer when she is filling her cart with groceries in Willy:s supermarket. The field of contemporary politics also seems to be structured around this fundamental issue. In today's capitalism, we are facing not one but at least two fundamental crises: there is the economic crisis that is currently manifesting itself in declining growth rates and increasing levels of debt in most western and even some eastern economies; and there is the ecological crisis that is manifesting itself in climate change, natural resource depletion and pollution. Transposing the Lenin joke into this context of a dual crisis of ecology and economy allows us to provide a crude mapping of the two-plus-one options that are currently available to mainstream politicians, while at the same time carving out the position of eco-analysis:

> A left-wing ecologist, a right-wing economist, a progressive liberal politician in the government of a small country in Northern Europe and Slavoj Žižek are asked whether they would prefer to save the planet or save capitalism. As expected, the left-wing ecologist answers, 'I would prefer to limit my consumption of capitalism's output of superfluous commodities in order to save the planet from global warming and other forms of ecological catastrophes.' The right-wing economist opts for capitalism, saying, 'I would prefer to maintain my current pattern of consumption thus propelling the growth needed to solve the economic crisis and save capitalism. Only the system of capitalism is efficient enough to provide us with resources required to solve the (alleged) ecological crisis and save the planet (if it is indeed in danger at all).' Now the progressive liberal politician in the government of a small country in Northern Europe says: 'I'd like to save both the planet and capitalism through technological innovations (sustain-

able energy, organic production of food, recycling, etc.) that allows capitalism to maintain its current volume of output and consumption but without the damaging effects to nature. In other words, the evolution of "green capitalism" gives me access to the decadent jouissance of both excessive consumption and the good conscience of knowing that I'm saving the planet.' Then Žižek answers: 'I would also prefer to save both the planet and capitalism.' Obviously, everyone is surprised by this answer. 'Why? Do you share with the progressive liberal politician in the government of a small country in Northern Europe the same decadent jouissance of capitalist consumption with a good conscience?' 'No,' Žižek explains: 'So I can tell the ecologists that I am trying to save capitalism, and tell the capitalists that I am trying to save the planet.' And then what do you do? 'Analyse, Analyse, Analyse!'

This fourth option is of course the position of eco-analysis. The point here is not so much that the dual crisis of ecology and economy provides us with the blueprint for a new form of society. The point is rather that, faced with the apparent choice between ecology and economy, we should refuse all of the three most apparent options: choose ecology over economy, choose economy over ecology, or find a third way that allows us to enjoy both at the same time. The 'Leninist' fourth option functions to break up and reconfigure the deadlock of the choice between ecology and economy. The challenge is to choose simultaneously ecology and economy but not because of their positive properties, their inherent fantasies of the good life, but rather because each of them offers the opportunity to refuse the other. By choosing both ecology and economy, we are able to refuse both. On the surface, the choice between ecology and economy presents itself as a choice between being against or for capitalism. This is why ecology is typically associated with the political left. We should be careful not to accept this premise all too readily. While it is true that ecology offers a platform from which certain aspects of capitalism may be criticized, there is still the danger that this particular platform is an inherent part of the problem rather than the solution. What if ecology is a way of being against capitalism, while still remaining within the fundamental structure of capitalism?

The refusal of identifying with any of the first three positions is also a refusal of immediate action. The real Žižek (and not only the imaginary Žižek conjured up in our rewriting of the joke) has proposed that, in the face of the impending global catastrophes of climate change and financial breakdown, the time is not for action but rather for thinking. The point is not so much that no action is needed to solve the problems of our time but rather that we should be open and honest enough to recognize that we actually do not know what kind of action is needed. As the fourth option breaks up the deadlock of the apparent choice between ecology and economy, it opens up a space for thinking. This is why we shall be referring to this as the position of eco-analysis. It is a position of critical analytical thinking.

What is the object of eco-analysis? As the observant reader will have already figured out from the above mention of Žižek, the concept of eco-analysis is a paraphrase of the concept of psychoanalysis. Methodologically speaking, the object of eco-analysis is the application of the method of psychoanalysis on another field. Substantially speaking, the object of eco-analysis is the 'eco'. Now, of course, this is anything but self-explanatory.

Given the apparent opposition between ecology and economy, it is ironic that the etymological roots of the two terms are closely related. The two words share the same prefix. 'Eco' is derived from the greek *oikos*, which means household, house or habitat. So if we go exclusively by the etymological definition, ecology, as well as economy, is concerned with the same object. Both are concerned with the place where life takes place. And even when we look at the ending of the words, the difference does not seem to be big. In ecology, 'logy' is derived from the greek *logos* that means 'word', 'speech', 'discourse' or even 'reason'. Of course, we also recognize the affinity to 'logic'. Along etymological lines, we may understand ecology as discourse and reasoning about the logic of places where life takes place. In economy, '-nomy' is derived from the Greek *nomos*, meaning 'law' or 'rules'. From a contemporary perspective, the notion of law is ambiguous. Law may refer to normative laws, such as legal or moral laws, but the word may also refer to laws embedded in the nature of things, for instance, the law of gravity. Nevertheless, we may summarize the etymological understanding of economy as the laws of places where life takes place.

The result of this sketchy etymological analysis shows that the difference between ecology and economy is merely that the first is concerned with the logic of habitats while the latter is concerned with the laws of habitats. If this were all we knew about ecology and economy, we would think that the two were very closely related and that knowledge in both fields was heavily influenced by the other, perhaps even to the point where they would conflate. However, few phenomena develop in strict accordance with their etymological roots, and their current meanings often point far beyond their initial definitions. This certainly applies to ecology and economy.

With Žižek, we may invoke the notion of parallax in order to unfold not only the relation between ecology and economy but also to uncover the underlying notion of 'eco', which seems to be the shared object of the two modes of inquiry. At the heart of Žižek's thinking, we find the threefold distinction between the real, the symbolic and the imaginary. This triad, that Žižek has of course inherited from Lacan, constitutes a distinction between different ontological domains. Whenever we want to understand concepts or phenomena such as subjectivity, language, love, politics, law, poker, democracy, money or the body within the framework of Žižek's thinking, we need to analyse how the concept of phenomenon is constituted in the interplay between these three domains. In the following, we are going to see how ecology and economy fit into the ontological triad of real, symbolic and imaginary. We shall start by seeing how the eco is real.

Throughout his oeuvre, Žižek offers a number of definitions of the real, as well as a number of analytical tools by which the real may be uncovered. At some points, the real is located in a positive existence beyond the sphere of symbolization. He defines the real as 'that which resists symbolization' and 'as the rock upon which every attempt at symbolization stumbles'.[2] At other points, the real is located in a negative existence, i.e. as merely a void or an aporia inherent in the symbolic order. Žižek states: 'the symbolic order itself, is…*barré*, crossed-out, by a fundamental impossibility, structured around an impossible/traumatic kernel, around a central lack.'[3] This lack is the real. The two ways of theorizing the real may immediately seem contradictory, and perhaps they are. Yet this contradiction is in itself a symptom of the impossibility of conceptualizing (symbolizing) the real. Žižek also provides a dynamic conceptualization of the real through the notion of parallax, which seems to combine the positive as well as the negative definitions of the real. Since the notion of parallax is key to our analysis of the eco as real, I shall be quoting Žižek at length on this issue:

> The Real is thus the disavowed X on account of which our vision of reality is anamorphically distorted; it is simultaneously the Thing to which direct access is not possible and the obstacle which prevents this direct access, the Thing which eludes our grasp and the distorting screen which makes us miss the Thing. More precisely, the Real is ultimately the very shift of perspective from the first standpoint to the second. Recall Adorno's well-known analysis of the antagonistic character of the notion of society: in a first approach, the split between the two notions of society (the Anglo-Saxon individualistic-nominalistic notion and the Durkheimian organicist notion of society as a totality which preexists individuals) seems irreducible; we seem to be dealing with a true Kantian antinomy which cannot be resolved via a higher 'dialectical synthesis,' and elevates society into an inaccessible

Thing-in-itself; in a second approach, however, we should merely take note of how this radical antinomy which seems to preclude our access to the Thing is already the Thing itself – the fundamental feature of today's society is the irreconcilable antagonism between Totality and the individual. This means that, ultimately, the status of the Real is purely parallactic and, as such, nonsubstantial: it has no substantial density in itself, it is just a gap between two points of perspective, perceptible only in the shift from the one to the other. The parallax Real is thus opposed to the standard (Lacanian) notion of the Real as that which 'always returns to its place' – as that which remains the same in all possible (symbolic) universes: the parallax Real is, rather, that which accounts for the very multiplicity of appearances of the same underlying Real – it is not the hard core which persists as the Same, but the hard bone of contention which pulverizes the sameness into the multitude of appearances. In a first move, the Real is the impossible hard core which we cannot confront directly, but only through the lenses of a multitude of symbolic fictions, virtual formations. In a second move, this very hard core is purely virtual, actually non-existent, an X which can be reconstructed only retroactively, from the multitude of symbolic formations which are 'all that there actually is.'[4]

In a superficial reading, the conflict between ecology and economy may seem to be a rivalry between two mutually exclusive understandings of the world in which we live. As we have already touched upon in the rewritten version of the Lenin joke, the ecological view of the world proposes that we live in accordance with nature by limiting our interference with the inherent balance of natural ecosystems, while the economic view is primarily concerned with optimizing the productive output of the world in order to maximize the satisfaction of human needs and wants. The object of contestation between the two perspectives is of course the eco, i.e. the place where we live. Should this habitat be viewed as a place where plants, animals and humans live in mutually beneficial harmony or should it rather be seen as a site of production and human consumption? Again, this is where we should resist the temptation of simply choosing the easy option of ecology over economy or even the cheap compromise between the two in the form of a sustainable economy that operates on the premises of ecology. This is not because these options aren't probably morally superior to the second option but because they constitute an analytical dead end. But of course this does not mean that we should choose the second option of simply giving in to the economic account of eco.

Starting from the bottom part of the quote, we may paraphrase Žižek to see how the parallax of the real pertains to the notion of the eco:

> In a first move, eco is the impossible hard core which we cannot confront directly, but only through the lenses of the symbolic fictions of ecology and economy. In a second move, this very hard core is purely virtual, actually non-existent, an X which can be reconstructed only retroactively from the multitude of symbolic formations of ecology and economy which are 'all that there actually is.'

The point is that ecology and economy are both symbolic accounts of the reality of the eco. Both accounts belong in the order of the symbolic and they each provide different systems of symbolization of the real of the eco. We shall be returning to the substantial claims of each of these systems. Neither ecology nor economy presents itself as a system of symbolization that is merely added to the existence of the eco as nature-in-itself. Instead, each of these symbolic formations claim to be recordings of the 'logy' and the 'nomy' that is already inherent in the real of the eco. Now, a crucial point in Žižek's philosophy is that the process of symbolization is never just a neutral recording of entities and events pre-existing in the order of the real. The process of symbolization is at the same time a production of the very reality that appears in the symbolic order. In other words, we may initially conceive of the real as just the undifferentiated, chaotic matter of being. This is perhaps captured by the interpretation of eco (*oikos*) as habitat that is simply the place where we live. However, through processes of symbolization, the real is rendered into the ordered and regulated reality in which we ordinarily find ourselves. *Oikos* is inscribed into the order of *logos* or *nomos*. According to Žižek, the operation of symbolization is not determined by qualities inherent in the objects of the real. On the contrary, certain paradigms of meaning and regularity are reproduced within the symbolic order on the basis of structures inherent in this order. In this respect, Žižek is in concurrence with mainstream social constructivist thinking, which we find also in figures such as Wittgenstein, Luhmann or Foucault.

If we observe here Žižek's distinction between the social *reality* and the *real*,[5] we see how the symbolization of the real is a social construction of reality. The mere undifferentiated Being of *oikos* or eco belongs in the order of the real. The real of the eco emerges as the reality of ecology or economy when it is incorporated into the symbolic order of economic signs that render entities and events meaningful in terms of their relations to other economic signs. This means that, in the first move, the undifferentiated being of the real disappears as it is transformed into the reality of the symbolic order. It becomes 'the impossible hard core which we cannot confront directly'.

Now the point where Žižek breaks away from conventional social constructivism is in his axiomatic insistence on the ultimate incompatibility between the symbolic and the real. In any operation of symbolization, there is an excess or a lack in the correspondence between the symbolic order and the real. This means that in the second move, the real re-emerges as something that is left out in the first process of symbolization. With the notion of parallax, Žižek even distances himself from Lacan in that the real is not merely some material common denominator that 'remains the same in all possible symbolic universes'. In turn, the parallax real is defined as 'that which accounts for the very multiplicity of appearances of the same underlying Real'. The parallax real can be discovered neither through the first, nor through the second, nor even from some third neutral perspective. The parallax real emerges only in the shift from the first to the second perspective.

It must be granted that Žižek's formulations here about the parallax real are rather enigmatic. The purpose of this chapter, as well as other parts of this book, is precisely to see how these formulations make sense in the context of ecology and economy. The hypothesis that we shall be proposing is that ultimately ecology is an account of the world without human subjects. In turn, economy is ultimately an account of the world where the human subject is reduced to an individual consumer that is then posed as the measure of every object in the world. In the shift between these two perspectives, the parallax real of the eco emerges as something that is simultaneously too objective and too subjective. Žižek's reference to Adorno may help illustrate this. Instead of engaging directly in the classic sociological dispute about whether society ultimately consists of a conglomerate of individuals or whether society has its own emergent properties that ultimately shape and regulate the behaviour of individuals, Adorno opts for the fourth position by proposing to see this conflict as an inherent property of society itself. (The route of the sociological third position not taken by Adorno is of course the form of theoretical synthesis found in Anthony Giddens's theory of structuration or Pierre Bourdieu's theory of the habitus as a 'structured and structurating structure'.) In the context of eco-analysis, the fourth Leninist option means that the truth about the eco lies neither in the ecological account, nor in the economic account. Still, this does not mean that they are both false. It means that the very tension between these two accounts is inherent in the very being of the eco. We should refuse both because this engages us in a negative dialectic that allows us to uncover the parallax real of the eco in the very shift between perspectives. This is the purpose of eco-analysis.

The logic of the Lenin joke serves to structure the course of the book. In part I, we shall begin our analysis by looking into ecology. If ecology is a particular symbolic order, which fantasies function to sustain this order? How does ecology symbolize the real? And what is the position of the subject in the ecological conception of nature? These are some of the questions guiding our eco-analytical exploration of ecology. Part of the analysis is a genealogy of the emergence of ecology as a distinct field of knowledge but still emphasis is put on the ontological properties of the constitution of ecology.

In part II, we proceed by analysing the economy. Where does value come from? How does economics conceive of the relation between price and value? What does it mean that the economy is growing? And what is the position of money in the economic conception of the world? Again the analysis begins with a genealogical exploration of the history of economics, which distinguishes between physiocrat economics, classical economics and neo-classical economics in terms of their respective ontologies of value. This leads into a critical examination of contemporary understandings of growth, knowledge, technology, consumption, capital, labour and production.

Part III is an engagement with contemporary ideas of so-called 'green growth'. How do contemporary policies of sustainability function to preserve an imperative of perpetual economic growth? Which are the fantasies inherent in the ideology of growth? How does our current mode of economic organization reinforce the imperative of growth? And what is the relation between money, debt and interest on the one hand and economic growth on the other? This part is organized around Žižek's concepts of need, desire and drive in order to map out three different dimensions of the propulsion towards economic growth that characterizes contemporary growth capitalism.

The book does not offer a catalogue of solutions to the ecological or the economic crisis. There are already plenty of other books offering this. Rather, it aims to shift the inquiry from: 'What shall we do?' to 'Why have we not already done it?' In order to properly address the challenges of our contemporary times of crisis, it is not enough to point out all the different reasons why perpetual economic growth is impossible and unsustainable. We need to understand how the idea of growth is deeply ingrained in the ideology as well as the organization of our society. And we need to understand why the idea of growth is so difficult to let go of. The purpose of the book is to open up the space for philosophical thinking about this issue.

Part I

The Logy of Eco

1

Balance of Nature

The paradox of radical ecology, which blames humanity for disturbing the natural homeostasis, is that, in it, a self-relating reversal of this logic of exclusion takes place: the 'excrement,' the destructive element which has to disappear so that the balance can be re-established, is ultimately *humanity itself*. As a result of its hubris, its will to dominate and exploit nature, humanity has become the stain in the picture of the natural idyll (as in those narratives in which ecological catastrophe is seen as the revenge of Mother Earth or Gaia for the wounds inflicted on her by humanity). Is this not the ultimate proof of the ideological nature of ecology? What this means is that there is nothing more distant from a truly radical ecology than the image of a pure idyllic nature cleansed of all human dirt. Perhaps, then, in order to break out of this logic, we should change the very coordinates of the relationship between humanity and pre-human nature: humanity *is* anti-nature, it *does* intervene in the natural cycle, disturbing or controlling it 'artificially,' postponing the inevitable degeneration, buying itself time. Nevertheless, as such, it is still part of nature, since 'there is no nature.' If Nature conceived as the balanced cycle of Life is a human fantasy, then humanity is (closest to) nature precisely when it brutally establishes its division from nature, imposes on it its own temporary, limited order, creating its own 'sphere' within the natural multiplicity.[1]

Often, the notions of ecology and economy appear to be in opposition or even in direct conflict. During the years of growth and relative prosperity immediately prior to the financial crisis of 2007–8, there was a growing concern about ecological issues such as global warming and pollution, and attempts were made to mobilize a political collaboration to address these issues on a global level. However, these efforts were immediately crushed with the onset of the crisis. Instead of ecological issues, political priorities were directed against economic issues related to growth and employment. Solving ecological problems were put on hold until allegedly more pressing economic issues had been dealt with. This was a clear example of ecology and economy being in opposition. Furthermore, it is generally the case that most of the ecological problems that we are facing are precisely caused by a one-sided focus on economic interests. The pursuit of monetary wealth, consumption opportunities and growth puts strain on natural ecosystems and threatens to throw them out of balance. In the often quoted words of one of the pioneers in ecological economics, E. F. Schumacher: 'Infinite growth of material consumption in a finite world is an impossibility.'

As we observe the trends and prognoses of global climate change, depletion of natural resources and extinction of animal species pointing towards future natural catastrophes, the most puzzling question is not: 'What shall we do?' By now, most people in the western world, who are also the ones doing most of the polluting, know full well what we should do: We should stop buying and using so much stuff that we do not need anyway. We should start eating beans and vegetables rather than meat from cows, pigs and other animals that demand several times more resources to generate similar amounts of nutrition. And we should stop transporting ourselves unnecessarily in cars and aeroplanes. In terms of informing people about the right ways of living an ecological lifestyle, the environmental revolution of the past four decades has been a huge success. We pretty much know what to do in order to save the planet. In turn, this only leads us to the real puzzle: why have we not already done it?

In the above quote, Žižek points to the ideological nature of ecology. While there are good reasons to concur with Žižek's observation, as we are going to see in the following, we may still ask whether the problem with ecology is that it is not ideological enough. If ideology is the phantasmatic frame that provides us with the coordinates of our desires, then the ideology of ecology seems to have had only marginal effects in terms of shifting us in the direction of an ecological lifestyle. Even if some progress has been made in terms of making consumers more aware of buying ecologically friendly products, such as biodynamic bananas or cars with low CO_2 emissions, this seems only to have been counterbalanced by increasing volumes of consumption overall.

But perhaps it is even misleading to think of ecology as an independent ideology trying to persuade us to live in greater harmony with the inherent balance of nature as opposed to the ideology of the economy propelling us towards more and more consumption. Perhaps ecology and economy are merely two sides of the same coin and thus part of the same coherent ideology. The challenge of eco-analysis in this connection is to keep one's eyes on the ball and not be swayed in the analysis of eco. In the following, we shall be looking into the history of ecological thinking in order to understand the genealogical relations between ecology and economy.

The Economy of Nature

A fundamental idea in ecological science, as well as in other forms of ecological thinking, is the notion of the 'balance of nature'. The notion of the balance of nature may be traced back to ancient Greek thinking.[2] It is derived from observations suggesting that nature is imbued with regulatory principles serving to secure stable populations of different species of animals that are balanced against each other. Herodotos (died c. 425 BC) made the following observation about the differential reproductive capabilities of predators and prey: 'The wisdom of divine Providence...has made all creatures prolific that are timid to eat, that they be not diminished from off the earth by being eaten up, whereas but few young are born to creatures cruel and baneful.'[3]

In the Middle Ages, the balance of nature was conceived in purely theological terms as just another other aspect of the providential vision of an almighty God. Concurrent with the Protestant Reformation, there is an emerging effort to found the notion of the balance of nature as a natural rather than a divine principle. This effort is of course in sync with the advent of modern thought found across a wide range of fields, such as science, arts, law, politics and, most importantly in the context of the present analysis, economy.[4] The differentiation of independent fields of thinking from the overall frame of theology is obviously a slow progression over several centuries. This is also the case with regards to ecology. As empirical data and methods in the fields that we have subsequently defined as zoology, botany, microbiology and so on improved, so there was also a gradual refinement of the balance-of-nature concept.

In 1713, William Derham was the first explicitly to use the word 'balance' in the context of ecology: 'The Balance of the Animal world is, throughout all Ages, kept even, and by a curious Harmony and just Proportion between the increase of all Animals, and the length of their lives, the World is through all Ages well, but not overstored.'[5] This line of thinking is taken one step further by Carl Linneaus. In

1759, he published the work *Oeconomia Naturae* which begins with the following axiomatic statement:

> To perpetuate the established course of nature in a continued series, the divine wisdom has thought fit, that all living creatures should constantly be employed in producing individuals, that all natural things should contribute and lend a helping hand towards preserving every species, and lastly that the death and destruction of one thing should always be subservient to the restitution of another.[6]

With the coining of the phrase 'economy of nature', Linneaus extends the balance of nature to include not only the animal world but also the domain of plants. Even though we still find in both Derham and Linneaus an element of divinity in the account of the balance of nature, it should be noted how there is also an emerging sense of the balanced perpetuation of nature as being a purpose in itself for which divinity provides the necessary conditions. The purpose of nature is not only to serve as venue for the display of divine omnipotence, generosity and foresight but nature also has a value in and of itself. As already indicated, the genealogy of ecology follows the same general pattern of a gradual decoupling from theology, spanning the Middle Ages through the Reformation and into modern times, that may also be traced in other fields of knowledge and thinking. Before moving on to see how ecology evolves into becoming a science in its own right, we shall see how the balance of nature is parallel to ideas found within the field of economics during the same period.

In 1759, the same year that Linneaus published his *Oeconomia Naturae*, Adam Smith published his book on *The Theory of Moral Sentiments*. In this work, Smith invokes for the first time the famous notion of 'the invisible hand' to account for the way that even the selfish behaviour of the rich landowner is part of an intricate division of labour whereby the products of society are distributed among all of its members.

> The produce of the soil maintains at all times nearly that number of inhabitants which it is capable of maintaining. The rich only select from the heap what is most precious and agreeable. They consume little more than the poor, and in spite of their natural selfishness and rapacity, though they mean only their own conveniency, though the sole end which they propose from the labours of all the thousands whom they employ be the gratification of their own vain and insatiable desires, they divide with the poor the produce of all their improvements. They are led by an invisible hand to make nearly the same distribution of the necessaries of life, which would have been made,

had the earth been divided into equal portions among all its inhabitants, and thus without intending it, without knowing it, advance the interest of the society, and afford means to the multiplication of the species. When Providence divided the earth among a few lordly masters, it neither forgot nor abandoned those who seemed to have been left out in the partition.[7]

While Linneaus provides an account of the economy of nature, one of the ambitions of Smith's work is perhaps rather to provide an account of the nature of economy. This ambition becomes even more spelt out with the publication of *The Wealth of Nations* in 1776. In this work, the notion of the invisible hand reappears more explicitly within the context of the market economy. This is the passage that is typically quoted to illustrate the point:

As every individual, therefore, endeavours as much as he can both to employ his capital in the support of domestic industry, and so to direct that industry that its produce may be of the greatest value; every individual necessarily labours to render the annual revenue of the society as great as he can. He generally, indeed, neither intends to promote the public interest, nor knows how much he is promoting it. By preferring the support of domestic to that of foreign industry, he intends only his own security; and by directing that industry in such a manner as its produce may be of the greatest value, he intends only his own gain, and he is in this, as in many other cases, led by an invisible hand to promote an end which was no part of his intention. Nor is it always the worse for the society that it was no part of it. By pursuing his own interest he frequently promotes that of the society more effectually than when he really intends to promote it.[8]

The parallel between the balance of nature and the invisible hand lies in the idea that every member and every part of the community is integrated into a system of mutual interdependence, where the natural behaviour of each individual, even when this behaviour is motivated by sheer survival or other individual desires, ultimately serves to benefit the reproduction of the whole. The difference between the two ideas lies of course in the scope of the unity under consideration. The simultaneous publication of Carl Linneaus's *Oeconomia Naturae* and Adam Smith's *The Theory of Moral Sentiments* provides an eminent genealogical example of the parallax of ecology and economy. Both works invoke the gaze of God, in the form of 'divine wisdom' and 'Providence' respectively, as the point of view from which the inherent balance in the order of the eco emanates. Still, the divine view of the world projects two different accounts of the ordering principles of the world. For Linneaus, the balance of nature applies to the community of animals and plants in a natural habitat. For Smith, the invisible hand applies to the capitalist market, where goods and services are produced and exchanged between human agents appearing as producers and consumers. In formal terms, the eco described by Linneaus is ultimately a world of objects, while the eco described by Smith is a world where everything is ultimately measured by the standards of human subjectivity.

The Split Eco

Rather than debating which of these two accounts of the world should be regarded as primary and superior to the other, we need to stay on the narrow path of eco-analysis and thus interpret the duality of the perspectives of Linneaus and Smith as perhaps a founding moment in the constitution of something that we might term 'the split eco'. This term is of course a paraphrase of Žižek's concept of the split subject that is sometimes also designated using Lacan's intricate system of notations by the symbol $. As the concept of the split subject provides us with another theoretical handle to understand the parallax of eco, we shall temporarily digress from the genealogy of ecology to present this concept.

The concept of the split subject refers to the subject's position as an intermediary between the order of the real and the order of the symbolic. Žižek explains: 'the subject is not directly included in the symbolic order: it is included as the very point at which signification breaks down. Sam Goldwyn's famous retort when he was confronted with an unacceptable business proposition – "Include me out!" – perfectly expresses this intermediate status of the subject's relationship to the symbolic order between direct inclusion and direct exclusion.'[9]

The notion of the split subject can be illustrated by the way we ordinarily think of ourselves as individuals. Individuality implies that something is a unity; it is *in*-dividable. All the time we refer to ourselves as such a unity: 'I don't like cheese', 'I live in an apartment', 'I owe you $500', and so on. We assume the existence of a self as a unitary origin of our feelings, preferences, opinions, and so forth. Now, Žižek starts from the complete opposite view of the subject. Subjectivity is not a unity but rather the product of an irreconcilable gap. A person is incorporated in the symbolic order as he or she attains a certain position in the social structure of society. The ritual of giving a name to a person signifies this incorporation. As the person is incorporated into the structure of the symbolic order, this structure also provides the conditions of possibility for the person's reflective self-conception. This is Foucault's point about the social constitution of subjectivity, which is now a central component of mainstream social constructivist theories of the self.

Žižek breaks away from these theories as he insists that the incorporation of a person into the symbolic order is always incomplete. The subject will indeed identify with the symbolic designations attributed to us in the symbolic order. I will identify myself as a male, Danish, father, taxpayer and so on. But at the same time my identification with these designations is marked by a sense of misrecognition. Even if it were possible to make a complete list of all my symbolic designations, I would still have the feeling that I am 'more than this'. This feeling is constitutive to Žižek's understanding of subjectivity.

The point here is not that there is beyond the sphere of our symbolic identity some external kernel of true subjectivity. This is the mistake made by neurobiological conceptions of subjectivity that locate subjectivity in the materiality of the brain. Just as the person confronted with the complete listing of his or her symbolic designations will insist on being 'more than this', so would a person confronted with a real-time scanner image of his or her brain. Subjectivity may neither be reduced to the order of the symbolic nor to the order of the real. Subjectivity is rather the gap separating the two.

In similar fashion, we should avoid thinking about the objectivity of the eco in terms of some real entity that is completely devoid of human symbolization or interference. The objectivity of the eco is rather the gap separating the objectivist account of Linneaus's ecology and the subjectivist account of Smith's economy. We may thus think of the eco as a kind of split object. The implications of this way of thinking about the eco shall be explored as we continue the genealogy.

The Polluting Animal

The nineteenth century saw a number of crucial turning points in the genealogy of ecological thinking. An intricate challenge to the balance-of-nature concept emerged as scholars of natural history, through the discovery of fossils, gradually reached the conclusion that some animal species had been exterminated. The preservation of species had hitherto been a major tenet in the balance-of-nature concept. Now the question was how to reconcile the idea of balance with the annihilation of entire species from the history of the world. Although this specific problem may not have been the explicit concern of many scholars working within the field, we can still interpret the two most important ideas of the nineteenth century in the genealogy of ecology as two different answers to the problem.

Of course, the most famous achievements in nineteenth-century biology are Darwin's ideas of the natural selection and the struggle for existence that was launched with the publication of *On the Origin of Species* in 1859. Although the image of nature as the scene of constant battles of life and death in the struggle for survival may not immediately invoke a sense of balance, we may still understand Darwin as the expression of a new notion of balance that had also emerged among many contemporary thinkers. The idea was that balance was not necessarily the same as stasis.

> A struggle for existence inevitably follows from the high rate at which all organic beings tend to increase. Every being, which during its natural lifetime produces several eggs or seeds, must suffer destruction during some period of its life, and during some season or occasional year, otherwise, on the principle of geometric increase, its numbers would quickly become so inordinately great that no country could support the product. Hence, as more individuals are produced than can possibly survive, there must in every case be a struggle for existence, either one individual with another of the same species, or with the individuals of distinct species, or with physical conditions of life.[10]

While the Darwinian world of a struggle for existence is indeed a world of constant instability, this instability is at the same time the source of a progressive and dynamic form of balance. Natural selection through competition ensures a constant evolution, where species as a whole adapt to a changing environment. When natural selection is extended to encompass not only the competition between different individuals of the same species of animals and plants but also the competition between entire species of animals and plants, it may serve as an explanation of the extinction of some species over the course of natural history without sacrificing the idea of an overruling evolutionary balance of nature. Darwin writes: 'Battle within battle must be continually recurring with varying success; and yet in the long-run the forces are so nicely balanced, that the face of nature remains for long periods uniform.'[11]

Another type of answer to the problem of the extinction of species can be traced back to an idea proposed by Georges-Louis Leclerc, Comte de Buffon, in the second half of the eighteenth century. Buffon raised the suspicion that man himself might have been the culprit in the extinction of large animals such as mammoths and mastodons. This suspicion may be taken to mark the introduction of man as a significant player in the evolution of natural history. Over the course of the nineteenth century, the role of man in the understanding and explanation of natural phenomena became more and more pronounced. Key insights into the interactions between different animal species as well as between animals and plants were derived from observations of the disturbance of balance in local natural habitats provoked by man's introduction of foreign species into the habitat.

Nineteenth-century Europe saw an immense acceleration in the evolution of coal- and steam-powered industrial capitalism. The powerful destructive forces of capitalism, identified by Marx among others, did not only work to transform economic, social and moral relations between people. Capitalism was not just about bourgeois exploitation of the working class. It was also a matter of the subsumption of more and more domains of nature under the regime of industrial production.[12] As the development of industrial capitalism unfolded, the role of man as a source of intervention and instability in the natural environment became more and more pronounced. In 1864, three years prior to Marx's publication of the first volume of *Capital*, naturalist George Perkins Marsh wrote in his book on *Man and Nature*: 'Man is everywhere a disturbing agent. Wherever he plants his foot, the harmonies of nature are turned to discords... [O]f all organic beings, man alone is to be regarded as essentially a destructive power, and...wields energies to resist which, nature...is wholly impotent.'[13] The actual phrase ecology (*Ökologie*) was coined around the same time, namely in 1866 by Ernst Haeckel with the publication of his work *Generelle Morphologie der Organismen*. Haeckel writes: 'By ecology we mean the body of knowledge concerning the economy of nature – the investigation of the total relations of the animal both to its inorganic and to its organic environment.'[14]

The significance of Haeckel's work was probably first and foremost to pave the way for ecology as an independent field of study, even though it would take another 40 years before ecology would develop into a scientific discipline in its own right.[15] Still, the coining and definition of the word may be taken to mark the ultimate departure of ecology from the domain of theology. The traces of divine providence, which were still found in Derham and Linneaus, are no longer present in the works of Darwin and Haeckel. In Darwin, as we have also already seen, the balance of nature is no longer sustained by divine wisdom and providence but rather by the natural selection and struggle for existence inherent in the evolution of natural history. He explains how competition between individuals within a species as well as between different species drives natural selection, which in turn governs nature towards the evolution of those individuals and species best fit to secure a continuous balance of nature. While competition may seem to be the source of immediate struggle and imbalance, it is in the longer run the source of balance. The competitive forces inherent in nature thus render divine providence unnecessary as a variable in the explanation of the balance of nature.

Within the field of economics, we can track a similar development. While the theological connotations in Smith's notion of the invisible hand are still open to interpretation, his successors in the so-called school of classical economics, including David Ricardo, John Stuart Mill and even Karl Marx, work to develop a concept of the market in which productivity and growth are facilitated by forces more genuinely inherent in the market itself. These are first and foremost the competitive forces operating to set prices and allocate production in accordance with the laws of efficiency, supply and demand. The following passage from Ricardo serves to illustrate this way of thinking:

> Under a system of perfectly free commerce, each country naturally devotes its capital and labour to such employments as are most beneficial to each. This pursuit of individual advantage is admirably connected with the universal good of the whole. By stimulating industry, by rewarding ingenuity, and by using most efficaciously the peculiar powers bestowed by nature, it distributes labour most effectively and most economically: while, by increasing the general mass of productions, it diffuses general benefit, and binds together, by one common tie of interest and intercourse, the universal society of nations throughout the civilized world. It is this principle which determines that wine shall be made in France and Portugal, that corn shall be grown in America and Poland, and that hardware and other goods shall be manufactured in England.[16]

In ecology, we see how the balance of nature is no longer sustained by God as an extra-natural being. Instead, the balance of nature is inscribed directly into the very functioning of nature. And in economics, we see how the 'universal good of the whole' is also not sustained by divine providence but rather by the 'pursuit of individual advantage' and thus the inherent forces of competition inscribed into the functioning of the market itself.

While Darwin's and especially Haeckel's work marks the disconnection of ecology from theology, the disconnection seems to highlight at the same time the re-emergence of a classic theological problem within the context of ecology, albeit in a different and naturalized form. This is the theodicy problem of evil, which is most prominently found in the writings of Leibniz around the turn of the seventeenth century.[17] Questions of theodicy address the apparent contradiction between an omnipotent and benevolent God on the one hand and the existence of evil and suffering in the world on the other. If God is good and has power over the world, why does he allow evil and suffering to exist? In the context of ecology, this problem emerges in another form: If nature is inherently balanced, why then do instabilities, imbalances and ruptures exist? As we have seen, Darwin and Buffon provide two apparently different answers to this naturalized version of the theodicy problem. Darwin extends the notion of the balance of nature to apply merely to the long run of natural history, which then allows for temporary imbalances and ruptures such as the extinction of individual species. Buffon, in turn, points to man as the foreign agent disturbing the otherwise balanced order of nature.

Despite their differences, both of these answers revolve around the same paradoxical status of man vis-à-vis nature. Darwin's concept of natural selection raises the question of whether the evolution of man himself is part of the balanced course of natural history. When human activity leads to the extinction of an animal, for instance the dodo, is this merely an expression of the fact that the dodo is simply not fit for its natural environment? Is Man himself merely the medium of the inherent force of the evolutionary balance of nature? And if we extend this to our contemporary situation with global warming, should we just view radical climate change caused by the expansion of capitalism as merely another chapter in the inevitable course of natural history? Probably Buffon and certainly Marsh would answer both of these questions in the negative, which of course has the philosophical implication that Man is an *un*-natural being.

When man is included in the gaze of ecology in order to explain certain phenomena such as the extinction of species or the destruction of natural habitats through pollution that obviously throws nature into a state of imbalance, this opens a curious question about the very ontology of man. If man has the capacity to disturb the inherent balance of nature, then he himself cannot be part of nature. While all other species by definition operate in concurrence with the logic of nature, because they *are* nature, the human being is the one species with the capacity to screw up nature. If pollution is the unnatural introduction of contaminants into the natural environment that cause adverse change, then the human being is distinguished from all other species by the capacity to pollute. All other species cannot by definition pollute because they are always already part of the natural order of things. The human being is the only *un*natural being. Or, it is in the nature of the human being to be *un*natural. Man is the polluting animal.

We see here the same paradox at work which we find in the Christian account of the fall of Man. Originally, Man was created to live in perfect harmony with the natural order of the Garden of Eden. However, as we all know by now, Adam and Eve transgressed the natural order of this paradise, causing them to be expelled. The enigmatic question is then of course whether the transgression emerged at the same time that Adam bit the apple or whether the transgression was already inscribed into the very human being of Adam and Eve. If the latter is true, then Adam and Eve were never in perfect harmony with the natural order. They were always already *un*natural. If the former is true, then why did Adam eat the apple in the first place? The story of Adam almost invariably invites the thought: 'If only Adam had not eaten the apple, we would have all been living carefree in the Garden of Eden.' Yet, this thought veils the underlying fact that, had Adam not eaten the apple, he would not have been Adam. He would have been nothing but a stupid monkey. It is the very *un*natural act of eating the apple that distinguishes Adam as a human being. Adam's fall was indeed a catastrophe, but without this catastrophe there would be no human beings.

The ecological perspective may look at nature as habitat in order to see how the lives of different species are balanced against each other. Yet the background against which the state of a given habitat is measured is the idea of nature still unspoiled by human interference. The ecological perspective takes as its starting point a world without human beings. If pollution is the introduction of contaminants into the natural environment that cause adverse change, then the very introduction of human beings into the world is itself the primordial act of pollution. This is comparable to the way that, once God had put Adam into the Garden of Eden, the garden was no longer a balanced harmonious paradise. The introduction of Adam was an introduction of imbalance into the otherwise balanced order of paradise in the same way that the introduction of human beings into nature is an introduction of imbalance into an otherwise balanced order.

These philosophical intricacies emerging from the introduction of Man into the context of ecology may be analysed as the subjective side of the notion of the split eco. As we have seen, the eco emerges in the parallax between ecology and economy. When we approach the nature of the eco exclusively from the perspective of ecology, this perspective invariably runs up against its own limitations. The claim that nature is ultimately governed by an inherent balance is challenged by the actual existence of imbalances and ruptures. This is the naturalized version of the theodicy problem, and the problem of Adam obviously revolves around the same inherent contradiction. And the only way of resolving this contradiction in ecology is by introducing Man as the unnatural being responsible for disturbing the balance. This Man is not just any kind of Man but a very particular kind of Man. This is Economic Man, the subject that turns the eco into a place of human production and consumption, even when it means the extinction of dodos and bisons.

In other words, the split eco is at the same time balanced and unbalanced, and the subject is simultaneously part of the eco and not part of the eco. This parallax nature of the eco thus implies the equally parallax nature of the split subject. We may recount here Žižek's formulation that 'the subject is not directly included in the symbolic order: it is included as the very point at which signification breaks down', which applies perfectly to the status of Man in the context of ecology. Man is not directly included in the ecological account of balanced nature. On the contrary, Man is included as the very point at which the ecological account of nature breaks down. Man is 'included out' of nature as he is introduced as the unnatural part of nature disturbing the balance. The subject of ecology is thus, paradoxically, the split subject that is simultaneously a product of natural evolution and an economic subject subsuming its habitat to the paradigm of production and consumption. In brief, the subject of ecology is *homo economicus*.

The Fantasy of Balance

Before we continue the genealogy of ecology into the twentieth century, we shall make another digression to introduce the third dimension of Žižek's ontological triad. This is the dimension of the imaginary. As we have already seen, the relation between the real and the symbolic is never a steady and coherent relation of representation. Even when the real has initially been symbolized and incorporated into the symbolic order, it eventually re-emerges in the form of something that has been left over or something that is missing. In psychoanalytical terms, the relation between the real and the symbolic is always already traumatic. The imaginary order is where this irreconcilable gap between the symbolic and the real is managed. Since there is no logically consistent solution to the problem of the relation between the real and the symbolic, the imaginary order has the form of fantasy. The symbolic order presents itself as the order of logic, calculation, rule of law, predictability, coherence, completeness etc. The order of the imaginary, on the contrary, takes the form of paradox, tautology, incoherence and ideology. In the order of the imaginary, we find a vague and often not fully articulated fantasy about a completed state of the symbolic order where contradictions and antagonisms have been overcome: '[T]he function of fantasy is to fill the opening in the Other, to conceal its inconsistency... Fantasy conceals the fact that the Other, the symbolic order, is structured around some traumatic impossibility, around something which cannot be symbolized.'[18]

Fantasy projects an image of the ontological gap between the symbolic and the real as merely a technical, practical and temporary problem, which may be overcome if only the proper measures are taken and incidental obstacles are cleared out of the way. As we have seen, the notion of the balance of nature is an inherent part of the constitution of ecology. In Žižek's terms, the balance of nature is the necessary fantasy sustaining the symbolic order of ecology. The 'traumatic impossibility' of ecology is perhaps the very symbolization of the real of the eco in a coherent system of logic. Harmony and rupture, stability and instability, life and death seem to co-exist in the order of the eco. The eco is one and the other at the same time. This ambivalent nature of the eco makes it difficult to subsume the functioning of the eco under one coherent logic. Furthermore, we may conceive of Man as both cause and solution to the traumatic impossibility of ecology. As the polluting animal, Man is the factor that always and everywhere disrupts the harmonious functioning of nature. But the disturbing intervention of Man is also what, in a second move, ultimately saves nature. By excluding Man from nature and seeing him as a foreign agent intruding into the domain of true nature, ecology is able to retain the image of nature as harmonious and balanced. If it were not for the disturbing intervention of the unnatural Man, nature would maintain itself in a state of balance.

Žižek often uses the terms 'imaginary', 'fantasy' and 'ideology' interchangeably. The function of the imaginary should, however, not be confused with the popular notion of ideology as a veil covering up the true state of reality. On the contrary, if we keep in mind the distinction between the real and reality, ideology is part of the very fabric of reality. In a key formulation, Žižek puts it this way:

> Ideology is not a dreamlike illusion that we build to escape insupportable reality; in its basic dimension it is a fantasy-construction which serves as a support for our 'reality' itself: an 'illusion' which structures

our effective, real social relations and thereby masks some insupportable, real, impossible kernel.... The function of ideology is not to offer us some point of escape from our reality but to offer us the social reality itself as an escape from some traumatic, real kernel.[19]

The imaginary may indeed serve to cover up an underlying traumatic split but the cover-up is an inherent part of the very functioning of reality. The imaginary is not a derivative form of ontological order, the neutralization of which would result in a state of truth. The truth does not reside somewhere behind or beyond the order of the imaginary but in the very imaginary interweaving of the real and the symbolic. This means that the notion of the balance of nature constitutes the ideological component in ecology. The 'traumatic, real kernel' from which ecology provides an escape is nature itself. Through the fantasy of balance, ecology provides an idea of nature as ultimately benevolent and harmonious. This is the reality of nature offered by ecology. However, this masks the fact that Man himself, even when he pollutes, exterminates entire species, or excretes CO_2 into the atmosphere, is an inherent part of the real of nature. We may also formulate this in terms of the distinction between the eco and ecology. The distinction between Man and nature is a product of the symbolization process of ecology. The distinction is not immediately inherent in the real of the eco. This is disturbing because it prevents us from pointing to the real of nature itself as a reason for why we should abstain from throwing waste into the oceans, shooting rhinos or extracting shale oil through fracking. Since Man is also ultimately a product of nature, it may equally well be argued that Man's excessive consumption, dominance of animals and exploitation of natural resources is an inherent part of nature. This is the traumatic real kernel that ecology allows us to escape through the ideological fantasy of the balance of nature. Retroactively, the fantasy serves to exclude everything that does not ultimately contribute to the balance of nature as being unnatural. In other words, whenever Man acts to disturb the perceived balance of nature, he is acting from a point outside of the domain of nature. We shall be looking further into this relation between Man, ecology and ideology in what follows.

2

Ecology Beyond Biology

Ecosystems and the Efficient Market

As we follow the progression of ecology into the twentieth century, we find that the paradox of the position of the human subject within the object of the eco has by no means been resolved. The development may rather be described as a kind of generalization of this problem. One of the most significant novelties in twentieth-century ecology is the development of the notion of ecosystems. This development can be traced back to the late nineteenth century with Stephen A. Forbes's seminal investigation of ecology in 'The Lake as a Microcosm' whose purpose was 'to study the system of natural interactions by which this mere collocation of plants and animals has been organized as a settled and prosperous community'.[1] The keyword in this formulation is 'community', by which Forbes means to emphasize the need to study ecology in terms of relations, interactions and interdependency both between different animal and plant species and between these species and their physical environment. We see here the contours of a methodological shift from the focus on individual species to the focus on emergent systems.

The now widely used term 'ecosystem' was not, however, coined until 1935 by botanist Arthur G. Tansley. In his reflections on the proper terms used to designate the basic units studied in ecology, he suggests that:

> the more fundamental conception is, as it seems to me, the whole *system* (in the sense of physics), including not only the organism-

complex, but also the whole complex of physical factors forming what we call the environment of the biome – the habitat factors in the widest sense. Though the organisms may claim our primary interest, when we are trying to think fundamentally we cannot separate them from their special environment, with which they form one physical system. It is the systems so formed which, from the point of view of the ecologist, are the basic units of nature on the face of the earth. Our natural human prejudices force us to consider the organisms (in the sense of the biologist) as the most important parts of these systems, but certainly the inorganic 'factors' are also parts – there could be no systems without them, and there is constant interchange of the most various kinds within each system, not only between the organisms but between the organic and the inorganic. These *ecosystems*, as we may call them, are of the most various kinds and sizes.[2]

Tansley's intention is not merely to promote the study of ecology in terms of interactions and exchanges between elements of a system along the lines already proposed by Forbes but also to emphasize the physical and chemical dimension of these interactions between the organic and the inorganic. This view was formulated in explicit opposition to two contemporary scholars, John Phillips and Frederic Clements, who had been using the term 'superorganism' as a metaphor to describe systems of cohabitation of different species in a given environment. Tansley was very critical about the metaphysical and almost religious connotations in the concept of the superorganism in so far as it was connected to a scientifically suspect form of holism.[3] The purpose of the concept of ecosystems is instead to ground ecology in the foundations of physics and chemistry, rather than metaphysics. While the concept of the superorganism tends to elevate the interactions and exchanges between different organisms and their environment into an emergent form of organism with a life of its own, Tansley's concept of ecosystems tends to move in the diametrically opposite direction by reducing even living organisms to their physical and chemical properties.

This approach of studying biological processes in physico-chemical terms, initially opened by Tansley, was further developed in the field of study known as biogeochemistry. One of the key proponents of this approach was Russian mineralogist Vladimir Vernadsky. Vernadsky replaces the concept of life (*bios*), which is the defining object of inquiry in *bio*-logy, with the concept of living matter:

> In biogeochemical processes...the totality of living beings – living matter, continues to play the basic role. It is characterized as the totality of all organisms, mathematically expressed as the totality of average

living organisms. Biogeochemistry studies, above all, the manifestation of the totality, not of the average indivisible unit. In the majority of the other biological sciences, we chiefly study the average indivisible unit; and, in the sciences of medicine and animal husbandry, the indivisible unit, individuality, or the single personality has been of outstanding significance during the past millennia.[4]

Although Vernadsky does indeed maintain that there is a difference between living matter and inert, non-living matter, the purpose of biogeochemistry is to provide a scientific rather than a philosophical account of this difference. Living and non-living matter is thus studied in terms of differences in their material composition, as well as in the way they absorb and transmit energy. Vernadsky conceptualizes this as a material-energetic distinction, which allows him to study the interactions between living and inert matter in physico-chemical terms. Such studies reveal the 'biogenic migration of atoms of the chemical elements, from the inert bodies of the biosphere into the living natural bodies and back again' that occurs through 'processes of respiration, feeding, and reproduction of living matter'.[5] In a sense, Vernadsky merely represents the next logical step from the concept of ecosystems but at the same time his conceptual displacement from life to living matter has the consequence of questioning the very discipline of biology as the proper domain for the study of ecosystems and ecology. If the difference between living bodies and inert bodies is ultimately a question of their physico-chemical composition, should they not just be studied with the theories and methods of the pure natural sciences of physics and chemistry?

There is another twist to Vernadsky's approach as he does in fact admit human beings a special place in his theory. Biogeochemistry is also a theory of evolution and according to Vernadsky:

> We are living in a brand new, bright geological epoch. Man, through his labor – and his conscious relationship to life – is transforming the envelope of the Earth – the geological region of life, the biosphere. Man is shifting it into a new geological state: Through his labor and his consciousness, the biosphere is in a process of transition to the noosphere. Man is creating new biogeochemical processes, which never existed before. The biogeochemical history of the chemical elements – a planetary phenomenon – is drastically changing. Enormous masses of new, free metals and their alloys are being created on Earth, for example, ones which never existed here before, such as aluminium, magnesium, and calcium. Plant and animal life are being changed and disturbed in the most drastic manner. New species and races are being created. The face of the Earth is changing profoundly. The stage of the noosphere is being created.[6]

But even with the notion of the noosphere, Vernadsky still reduces human life to a matter of chemistry. The way that the emergence of human life is registered in the history of natural evolution is through the physico-chemical impact of human activity on the biosphere. Human life does not constitute a new ontological dimension in the world. It merely creates new combinations of existing matter in the world. In the earlier accounts of ecology by Haeckel and Darwin, we encountered the paradoxical relation between man and the balanced order of nature. This paradox is anything but solved by Vernadsky. In fact, we may argue that Vernadsky merely takes the implications of this paradox even further.

The problem of the relation between Man and the balance of nature emerges in Vernadsky's argument as the problem of whether life is anything but a particular molecular formation. This is also known as the problem of emergentism. If we go along with the proposition that the difference between living and non-living bodies is ultimately a matter of physics and chemistry, then we are yet again confronted with the theodicy problem of explaining why the inherent balance of nature does not seem to extend to the kind of living bodies also known as humans. Either there is discontinuity between nature and man or nature is not balanced in the first place. But Vernadsky's biogeochemistry has a further and perhaps more radical implication as it may lead us to question the very value of human life. If living matter, including human beings, is ultimately just a particular configuration of energy and molecules, then why should anyone be concerned by the prospect of future disturbances of the balance of the biosphere leading to natural catastrophes and annihilation of life? As the concept of life is generalized into the concept of living matter, we lose the perspective of the individual human being against which the value of life may be measured. The concept of the noosphere also has the implication that the presence of human life on the planet may be merely a temporary stage in the history of natural evolution. Just like the dinosaurs had their five minutes of fame at some point in the course of evolution, so human beings shall also eventually perish and become only a brief chapter in the history of the planet. In this light, the destruction of those features of nature necessary for human well-being and survival may simply be part of the natural course of evolutionary history. This also means that any attempt to turn current trends around by limiting the burning of fossil fuel, recycling rather than disposing waste, banning different kinds of pollution and so on are actually efforts that work against the inherent logic of nature.

While this perspective may be objectively true, it is at the same time existentially false because it fails to take into account that there is no point from which the past and future history of the planet can be observed. We cannot translate such a view of the world into the lived experience of a subject. Even Vernadsky himself is nothing but a human subject that happens to be thrown into existence at a certain point in time. What is emphatically absent from any ecology derived from Vernadsky's biogeochemistry is precisely the figure that is the centre-point in economics: the individual human being. The individual human being is not only the polluting animal capable of screwing up nature. The individual human being is also the concerned animal capable of worrying and caring about whether nature is screwed up. Vernadsky's concept of the noosphere may have the purpose of urging us that we, the human race, are radically responsible for the continued functioning and reproduction of the biosphere. At the same time, the biogeochemical account of the world has the, perhaps unintended, consequence of stripping away any convincing argument as to why we should in fact care about the future of the biosphere.

Concurrent with the development of the concept of ecosystems within ecology over the course of the twentieth century, we find a comparable trend in economics. With the development of neo-classical economics, we see how the discipline narrows in on the study of markets as self-sustaining entities. This is most clearly visible in the development of neo-classical finance theory, which can be traced back to Louis Bachelier's *Theory of Speculation* published in 1900.[7] In this school of thought, the basic idea is that the interaction and competition between speculators and investors in financial markets function to generate prices in the market that are constantly representative of the available knowledge of the underlying assets traded in the market. In the ideal market, it is thus impossible to make money on speculation as any knowledge about the future development of prices is always already incorporated into the present prices. In practice, several of the key figures in the academic development of this theory have subsequently moved on to Wall Street to make fortunes from their knowledge. The standard account of this idea of markets is formulated by Eugene Fama in 1970 and is also known as the efficient-market hypothesis:

> THE PRIMARY ROLE of the capital market is allocation of ownership of the economy's capital stock. In general terms, the ideal is a market in which prices provide accurate signals for resource allocation: that is, a market in which firms can make production-investment decisions, and investors can choose among the securities that represent ownership of firms' activities under the assumption that security prices

at any time 'fully reflect' all available information. A market in which prices always 'fully reflect' available information is called 'efficient'.[8]

The efficient market is in other words imbued with the same kind of intrinsic balancing mechanisms that we find in ecosystems. The efficient market is a system capable of responding to external input while at the same time maintaining its own homeostasis.

At around the same time as Fama's formulation of the influential efficient-market hypothesis in finance theory, chemist James Lovelock formulated another hypothesis that has become equally influential within the field of ecology. Lovelock included not only plants, animals and abiotic material but also the entire atmosphere in the notion of one big global ecosystem. The atmosphere and the abiotic environment not only support life on the planet, but living matter itself has a significant impact on the chemical composition of the atmosphere. Lovelock thus proposes to understand Earth as one single living entity where everything is connected to everything and all components collaborate in sustaining homeostasis. This image of the world is stated as the Gaia hypothesis, suggesting that 'the total ensemble of living organisms which constitute the biosphere can act as a single entity to regulate chemical composition, surface pH and possibly also climate. The notion of the biosphere as an active adaptive control system able to maintain the earth in homeostasis we are calling the "Gaia" hypothesis.'[9] Although Lovelock's notion of Gaia is far from uncontroversial in the academic world of ecology, the idea has been very formative in popular thinking about ecology. We see this first and foremost in contemporary controversies about climate change. What is at stake in these controversies is the relation between all of humanity and the balance of the entire planet.

Brundtland in Samarra

A favourite cover image on books about ecology and environmental issues is a photograph of Earth seen from space. The first and most famous of this kind of image was taken from the Apollo 17 flight in December 1972 (Figure 2.1). The photo is said to have played a significant role in the awakening of global awareness about climate and environmental issues because it makes Earth appear lonely, vulnerable and in need of protection. The image is thus alleged to inspire emotions of responsibility and care in the observer. In this sense, the Apollo 17 image is significant in the development of ecological thinking. Ecology today is not first and foremost a specific academic

Figure 2.1 Earth from Apollo 17

research discipline. Over the course of the past four or five decades, ecology implies a set of political beliefs that aim to take into account the preservation and restoration of nature. While ecology of course is still also a field of research, it is today a much more general and popular way of thinking about the relation between man and nature. We shall conclude this genealogy of ecology by looking at this ecologicalization of politics.

In 1987, the UN published the report *Our Common Future*, also sometimes referred to as the Brundtland Report after the Norwegian Gro Harlem Brundtland who headed the commission responsible for it. The report was a continuation of the initiative that had been launched in 1972 with the United Nations Conference on the Human Environment in Stockholm. The purpose of the Brundtland Report was to set 'a global agenda for change' that put sustainability and environmental issues on the agenda of international politics and to further global collaboration in addressing these issues. The first four paragraphs of the report provide a thorough basis for illuminating the ambiguities and deadlocks of the contemporary politics of ecology. I shall thus quote these paragraphs at length:

1. In the middle of the twentieth century, we saw our planet from space for the first time. Historians may eventually find that this vision had a greater impact on thought than did the Copernican revolution of the sixteenth century, which upset the human self-image by revealing that the Earth is not the centre of the universe. From space, we see a small and fragile ball dominated not by human activity and edifice but by a pattern of clouds, oceans, greenery, and soils. Humanity's inability to fit its activities into that pattern is changing planetary systems, fundamentally. Many such changes are accompanied by life-threatening hazards. This new reality, from which there is no escape, must be recognized – and managed.
2. Fortunately, this new reality coincides with more positive developments new to this century. We can move information and goods faster around the globe than ever before; we can produce more food and more goods with less investment of resources; our technology and science gives us at least, the potential to look deeper into and better understand natural systems. From space, we can see and study the Earth as an organism whose health depends on the health of all its parts. We have the power to reconcile human affairs with natural laws and to thrive in the process. In this our cultural and spiritual heritages can reinforce our economic interests and survival imperatives.
3. This Commission believes that people can build a future that is more prosperous, more just, and more secure. Our report, *Our Common Future*, is not a prediction of ever increasing environmental decay, poverty, and hardship in an ever more polluted world among ever decreasing resources. We see instead the possibility for a new era of economic growth, one that must be based on policies that sustain and expand the environmental resource base. And we believe such growth to be absolutely essential to relieve the great poverty that is deepening in much of the developing world.
4. But the Commission's hope for the future is conditional on decisive political action now to begin managing environmental resources to ensure both sustainable human progress and human survival. We are not forecasting a future; we are serving a notice – an urgent notice based on the latest and best scientific evidence – that the time has come to take the decisions needed to secure the resources to sustain this and coming generations. We do not offer a detailed blueprint for action, but instead a pathway by which the peoples of the world may enlarge their spheres of cooperation.[10]

Reading this report today, more than 25 years after its publication, one cannot help noticing how little seems to have changed. The report points to the exact same issues of pollution, depletion of natural resources, climate change, extinction of species and disruption of ecosystems that are on the agenda today. And the report has the same urgency of tone found in similar texts written today: We are facing serious and impending threats but there is still hope if we get together and start acting now. On the one hand, we can see the report as a huge success in so far as it has achieved its goal of putting sustainable development and environmental issues on the agenda in a way in which we are still discussing them. On the other hand, the report reads in hindsight as a testimony to the inability of our alleged global community of world citizens to actually change the course of the planet's development. We have known for several years what we need to do. And yet we have failed to do it.

From an analytical perspective, it is also interesting to note how the report starts out by invoking the image of the globe seen from space. Now, if it were actually true that the Apollo 17 image sparked a new sense of responsibility and care towards this 'small and fragile ball', then why was the publication of this image followed by four decades of global impotence in terms of reversing the trends of environmental degradation? Perhaps the meaning of the image is not so unambiguous after all.

The Apollo image seems to invoke two contradictory sentiments simultaneously. There is indeed the aforementioned feeling of radical responsibility. We must save the planet! But at the same time the image perhaps also holds the answer to the contemporary question of why global awareness of climate and environmental issues has only been realized to a very limited extent in global action. While the object in the Apollo photo is obviously Earth, the image does not tell us who the observer actually is. Even though we know very well that the picture was taken by the Apollo astronauts, it confronts us with a gaze on Earth with which it is almost impossible to identify. The gaze that is staged in the photo, looking down on Earth, is divine rather than human. It may very well be that the image of a lonely and vulnerable Earth evokes feelings of care and responsibility, but to act on these emotions requires you to be a god rather than a mere human being. The human gaze does not view the world as a globe that is blue and green. In the individual human gaze, the world is flat. In some places it is blue and green, but it is also black, grey, paved, built over and full of other people. The purpose of invoking the Apollo image in debates about ecology and sustainability is of course to create global awareness and responsibility in the hope that people are able to translate this into concrete action in a world in which the earth is predominantly flat. But perhaps the effect is the exact opposite. No person is able to save the circular earth floating in space. Only God can do that. Rather than being a call for global action, the image of the globe is an invitation for passivity, false hope and perhaps cynicism.

In the quotation above, the impact of seeing the earth from space is compared to Copernicus' discovery in the sixteenth century that the sun rather than the earth is the centre of the solar system. But what exactly is this impact? The most immediate interpretation is of course to suggest that the vision of the planet from space is yet another step in the decentring of the ontology of Earth as well as of humanity. Not only is the earth, as already discovered by Copernicus, not the centre of the solar system, but even our solar system is merely a small and insignificant grain of dust in the great complex of the entire universe. However, it seems equally possible to suggest the opposite interpretation – that the view from Apollo constitutes the reversal of the Copernican revolution. The Copernican revolution had an impact not only on the narrow field of astronomy but also on human thought more generally as it collided with predominant theological ideas about the privileged position of the earth within the universe. The Copernican revolution thus constituted a milestone in the process of secularization that is a key element in the evolution of modernity. In so far as the Apollo image invokes the feeling that there is some divine power looking down on the earth from above, the vision may have the effect of reconstructing the relation between Man and God after the traumatic event of the Copernican revolution. Indeed, the earth may not be the centre of the solar system, let alone the centre of the universe; however, this does not mean that we have been forgotten by God and that He does not still have a keen interest in us.

At the end of the quote from the UN report, we find the formulation: 'This new reality, from which there is no escape, must be recognized – and managed.' This formulation ties in with Žižek's definition of ideology touched on above: 'The function of ideology is not to offer us some point of escape from our reality but to offer us the social reality itself as an escape from some traumatic, real kernel.' The domain of ideology is exactly the domain of the imaginary, and the ideological function of the Apollo image is precisely to offer us reality itself in a form that we may be able to recognize and manage. The ambivalent nature of the Apollo image allows us to recognize, manage and reconcile the feeling of radical responsibility for the ecological state of the planet with the feeling of utter impotence in terms of changing the course of planetary degradation.

As we have seen throughout the preceding analyses, ecology's view of the world is permeated by contradictions and paradoxes. The Apollo photo seems to provide a phantasmatic image of the earth where these traumatic features of the real of the eco are either reconciled or eliminated. Seen from space, the earth appears to be in perfect balance and harmony. And perhaps more remarkably, there seem to be no human beings. In this way, the Apollo photo is an image of the fantasy of ecology, where nature is in balance with itself and the polluting animal has been eliminated from the equation.

Reading through all four paragraphs, we find a particular narrative structure that is emblematic of political texts about global ecological issues. The first paragraph establishes a state of impending catastrophe and urgency. We are facing serious problems. The second paragraph proceeds to modify this apocalyptic vision by pointing to already existing positive developments. The very same features of modernity – global division of labour, economic growth, technological innovations and scientific discoveries – that have caused the current state of ecological crisis also hold the seeds of possible solutions to the crisis. In the third paragraph, the agency of 'people' is instituted as the collective subject with the capacity as well as the responsibility for bridging the gap between possibility and reality. 'We' are the makers of our own 'common future'. Proper action is not only going to save an already precarious status quo but will even reward us with more prosperity, justice and security in a new era of economic growth. The fourth paragraph continues along similar lines by stressing the urgency of the situation while at the same time pointing to the need for 'decisive political action'. While it is ultimately up to the people to act, established politicians cannot afford to sit on their hands while waiting for change to come from below. They must also act now.

The narrative structure of the four paragraphs may be compared to W. Somerset Maugham's beautiful fable of 'The Appointment in Samarra', which Žižek uses to illustrate the dialectics involved in the constitution of subjectivity:

> DEATH: There was a merchant in Baghdad who sent his servant to market to buy provisions and in a little while the servant came back, white and trembling, and said, Master, just now when I was in the marketplace I was jostled by a woman in the crowd and when I turned I saw it was Death that jostled me. She looked at me and made a threatening gesture; now, lend me your horse, and I will ride away from this city and avoid my fate. I will go to Samarra and there death will not find me. The merchant lent him his horse, and the servant mounted it, and he dug his spurs in its flanks and as fast as the horse could gallop he went. Then the merchant went down to the marketplace and he saw me standing in the crowd and he came to me and said, Why did you make a threatening gesture to my servant when you saw him this morning? That was not a threatening gesture, I said, it was only a start of surprise. I was astonished to see him in Baghdad, for I had an appointment with him tonight in Samarra.[11]

At least since the publication of the report *The Limits to Growth* in 1972,[12] which used computer modelling to project five major economic and ecological trends into the future: accelerating industrialization, rapid population growth, widespread malnutrition, depletion of non-renewable resources and a deteriorating environment, political debates about ecological issues have been informed by the prospect of future environmental catastrophes of a magnitude threatening the very foundation of civilization. It is the same kind of threat that is invoked in the Brundtland Report. The way the problem is formulated suggests that the danger is not that if we do something wrong, we shall be struck by catastrophe. Rather, the prospects of future catastrophe are merely projections of mechanisms inherent in our existing patterns of production, consumption, reproduction and so on. This means that catastrophes occur if we do not do anything radical to change the course of history.

In this way, the experience of reading reports like *The Limits to Growth*, *Our Common Future* or of watching Al Gore's 2006 documentary *An Inconvenient Truth*, which follows a similar narrative structure, is comparable to the experience of the servant encountering Death in the marketplace. The servant is scared, 'white and trembling', because he believes he is somehow destined to die. Equally, the projection of existing trends of pollution and resource depletion into the future gives the impression that deadly collapse is already inscribed into the destiny of our current system of capitalism. Now the shift between the first and the second paragraph in the Brundtland quote resembles the servant's paradoxical move. On the one hand, he believes that his own future is already predetermined and the encounter with the old woman is thus an omen of his forthcoming death. On the other hand, he also believes that he is capable of avoiding his own predetermined destiny. He is able to save himself by riding away to Samarra. In the second paragraph, the Brundtland Report gives us the same hope that we shall be able to escape our own destiny as it is inscribed in the existing order of capitalism. The people of the planet do not need to sit around in Baghdad awaiting their own self-destruction. We can change the course of our planetary development and choose to go to Samarra instead, where the death of capitalism will not find us. However, the formulations of the third paragraph may suggest that the report is leading us into the same trap as the poor servant.

The beauty in Somerset Maugham's fable lies of course in the way that the servant unintentionally cooperates in the realization of his own destiny. It is only through his own resistance that the predetermination of death is brought to a conclusion. Žižek uses the fable to illustrate the relation between the big Other and the constitution of the split subject. The servant represents the subject, who is reluctant to accept and identify completely with his symbolic mandate as determined by the logic of the symbolic order. Just as the marketplace is a domain for the circulation and exchange of commodities as well as human interaction, so the symbolic order is the domain for the circulation and exchange of signs constituting an ordering of subjects as well as of objects. Coming into being as a subject is connected with integration into the existing symbolic order. The subject is thus ascribed a certain social destiny. As we have already touched upon, the point where Žižek departs from more conventional social constructivist theories of subjectivity and identity is in his insistence that the subsumption of the subject in the symbolic order is never complete. Just as the real is 'that which resists symbolization', so we also find in the constitution of the subject an element of the real as the subject resists full identification with the symbolic mandate. The subject may recognize its position in the symbolic order but at the same time there is an inherent insistence that 'I am more than this.' The servant's act of riding off to Samarra to avoid Death represents this insistence that 'I am more than this.' The servant insists that he is more than his own destiny. Paradoxically, and this is where Žižek uses the fable to illustrate the function of ideology, resistance against the symbolic mandate is not necessarily a resistance against ideology. In fact, this very resistance may be already inscribed in the functioning of ideology. Within the context of capitalism, this means that some forms of resistance against capitalism may ultimately serve to reproduce and consolidate the system.

In the third paragraph, the UN report projects the image that 'people' do not just have the capacity to prevent the course of history culminating in ecological and economic catastrophe. We also have the potential to enter into a 'new era of economic growth'. It is hard to dislike the image of a future with more of all the things we already like: more prosperity, more justice, more security, more economic growth – which is at the same time also 'based on policies that sustain and expand the environmental resource base'. Still, we need to ask ourselves if such an imaginary fantasy does not work to maintain and perpetuate the very same structures of capitalist production and consumption that are causing the increasing ecological problems. Of course, we are here coming back to the fundamental question of whether ecological sustainability is at all possible within the structure of capitalism as such, or whether ecological sustainability can only be achieved through a revolution altering the fundamental coordinates of the economic system. The logic inscribed into the fable of 'The Appointment in Samarra' would suggest that even efforts to avoid ecological catastrophe by taking the path of 'green capitalism', that is by developing environment friendly products and sustainable forms of production, fail to question the fundamental imperatives of economic growth inherent in capitalism and therefore also fail to provide a true alternative capable of countering the destructive forces of capitalism.

Along these lines, it is also worth remarking the ending of the Al Gore documentary. Having spent the main part of the film painting a very gloomy picture of a planet heading towards global disaster, Gore ends with a number of quite modest calls for action to the viewer: change to energy-saving light bulbs; take the bike or walk, rather than driving; recycle your waste; eat less meat. And support an environmental group. And so on. While all of these suggestions are good and sympathetic, they still seem strangely out of proportion to the challenges and problems demonstrated by the documentary. And of course all of them are well within the boundaries of the typical lifestyle of consumer capitalism. The ending of the film thus seems to undermine the overall message: the whole of human civilization is heading towards self-destruction, yet this is nothing we can't handle with a few minor adjustments.

A very crude reading of the Brundtland quote along the lines of 'The Appointment in Samarra' would say that the ultimate ideological function of the report is to save the capitalist imperative of perpetual economic growth. Even if such a reading is a misrepresentation of many of the points and intentions of the report, it may still illuminate some of the elements that stand in the way of actually changing the trends of ecological destruction. In this respect, the last sentence of the third paragraph is worth noticing. The report not only invokes the possibility of a 'new era of economic growth', the necessity of economic growth is furthermore justified by the need to relieve poverty: 'we believe such growth to be absolutely essential to relieve the great poverty that is deepening in much of the developing world.' Who can argue against the need to feed the starving children of Africa? And yet perhaps this commitment to relieve poverty through continued economic growth is precisely what stands in the way of actually transforming our global modes of production and consumption beyond superficial adjustments.

The fourth paragraph stresses the need for immediate action. Even though projections for the future may appear pessimistic, this should not lead us to give up. There is still the opportunity to turn things around if only decisive political action is taken now. One philosophical lesson from the fable about the unfortunate servant in Baghdad is that an act, which we immediately perceive as an expression of free will resisting social predetermination, may turn out to be nothing but an element in the reproduction of social reality. The subject ultimately ends up being a medium for the big Other. On this note, we may reflect whether the kind of action suggested by the Brundtland Report is perhaps the kind of action that only functions to sustain the existing social order and thus fails to bring about the change needed to curb ecological deterioration of the planet. In part III, we shall return to the Brundtland Report to further explore such a claim.

'We Aren't the World!'

Nothing is easier than subjecting a political text such as the Brundtland Report to a philosophical deconstruction thus exposing its contradictions and implicit assumptions. Of course, this does not mean that the text nor the authors of the text are in any way stupid. In fact, the text contains a number of profound philosophical insights. Chapter 1, 'A Threatened Future', begins with the following formulation: 'The Earth is one but the world is not.'[13] This formulation, which looks like something out of the late Heidegger, captures very well a fundamental condition of the eco that we have been exploring in the preceding eco-analysis. The earth is the ultimate precondition for human existence. The report invokes the concept of the biosphere to capture this point as it continues from the preceding formulation: 'We all depend on one biosphere for sustaining our lives.'[14] But, curiously, the earth is both acutely present everywhere and all of the time, while at the same time it remains inaccessible. What we see when we look out the window is not the earth but the world. We never experience the totality of the earth. All we ever see, feel, smell and hear are small particular parcels of the world. A fundamental point in Žižek is that experience is enabled only through the framework of the symbolic order. This means that we have no direct access to the real. Žižek himself makes a similar point using the distinction between 'the scientific-technological' account of ecological catastrophe and the common-sense perception of the life-world:

> Disbelief in an ecological catastrophe cannot be attributed simply to our brain-washing by scientific ideology that leads us to dismiss our gut sense that tells us something is fundamentally wrong with the scientific-technological attitude. The problem is much deeper; it lies in

the unreliability of our common sense itself, which, habituated as it is to our ordinary life-world, finds it difficult really to accept that the flow of everyday reality can be perturbed. Our attitude here is that of the fetishist split: 'I know that global warming is a threat to the entire ecosystem, but I cannot really believe it. It is enough to look at the environs to which my mind is wired: the green grass and trees, the whistle of the wind, the rising of the sun... can one really imagine that all this will be disturbed? You talk about the ozone hole, but no matter how much I look into the sky, I don't see it – all I see is the same sky, blue or grey!' Thus the problem is that we can rely neither on scientific mind nor on our own common sense – they both mutually reinforce each other's blindness. The scientific mind advocates a cold, objective appraisal of dangers and risks, while no such appraisal is actually possible; common sense finds it hard to accept that a catastrophe can really occur.[15]

Žižek's distinction between the real and reality applies to the distinction between earth and world. The world is the reality in which we live among more or less recognizable objects and subjects to which different kinds of meaning are attached. The world is reality experienced from particular subjective positions within the world. The earth is the real that is both omnipresent and inaccessible. The earth is the totality that cannot even be divided into a finite number of particular experiences. Not even the Apollo photo allows us to experience the earth, although this is of course what constitutes the phantasmatic excess of the image. In turn, the Apollo image functions to veil the traumatic fact that the earth can never be experienced from within the world.

The relation between the earth and the world may be further elaborated by returning to Žižek's notion of the parallax real and by rewriting his definition to fit our particular purpose:

In a first move, the Earth is the impossible hard core of the Real which we cannot confront directly, but only through the lenses of a multitude of symbolically mediated experiences of the world. In a second move, this very hard core of the Earth is purely virtual, actually non-existent, an X which can be reconstructed only retroactively, from the multitude of symbolic formations of the world which are 'all that there actually is'.

First, the earth is the outer boundary of human existence that we cannot experience directly in its totality since we are always already limited by our embeddedness in the world. And then, secondly, the earth emerges merely as a phantasmatic image that has been reconstructed by a series of particular experiences of the world from within the world. The earth is one totality and the world is a collection of many particulars that do not add up to one total. The discrepancy between the earth, which is one, and the world, which is not, expresses the fundamental condition of the split eco. The eco is precisely split between this being-one-earth and being-many-worlds. This split is at the heart of the traumatic constitution of the eco.

Now the place where eco-analysis departs from the assumptions and aspirations of the Brundtland Report lies in the concept of trauma underlying the formulation: 'The earth is one but the world is not.' A general thread in the political rhetoric on global ecological issues that weaves through the Brundtland Report, as well as the growing number of similar declarations that have followed in its wake, is the idea that the world could become one. The title of the Brundtland Report, *Our Common Future*, invokes the fantasy that, even though the world is presently 'not one', it may eventually become one in the realization that we all share a common future. The report thus expresses the same hope as the song *We are the World* that was Michael Jackson's and Lionel Ritchie's famous contribution to the *Band Aid* campaign in 1985. The message of the song is that change will only come when the world stands together as one.

Both Brundtland and Jackson and Ritchie build on an underlying notion of trauma as something that can be resolved. The trauma is that the world is not one and it should be resolved by making the world come together as one. Now sticking to the formulation of psychoanalysis found in Žižek, eco-analysis would rather accept trauma as an ontological condition of the constitution of the subject as well as the eco. The split eco is not a contingent situation that has come about through some unfortunate event with traumatic consequences, and it makes no sense to attempt to eliminate the split through some form of reparation. In turn, the traumatic split between the earth and the world is an inherent feature of the eco. It is relevant to recapture how eco (*oikos*) is the place where life takes place. This is why the eco cannot simply be equated with the earth. The inclusion of life in the eco invariably brings with it the notion of the world and thus also the split between the earth and the world.

In his definition of the parallactic real, which was quoted at length in the introduction, Žižek makes reference to Adorno's analysis of the antagonistic character of society, which is split between 'the Anglo-Saxon individualistic-nominalistic' and 'the Durkheimian organicist notion of society'. Žižek shares with Adorno the point that society should not be conceived as either one or the other. Instead, he concludes, 'the fundamental feature of today's society is the irreconcilable antagonism between Totality and the individual'. The same point applies to the eco and the distinction between the earth and the world. The eco is neither the one totality of the earth nor is it the sum of manifold subjective views of the world. Indeed, the fundamental feature of the eco is the irreconcilable and traumatic antagonism between the (one) earth and the (many) world(s). Attempts to resolve this trauma are thus bound to fail as they come up against the ontological constitution of the eco.

As we look at the outcome of the Brundtland Report, together with the subsequent series of similar declarations formulated by international bodies such as the 1992 Rio Declaration *Agenda21*, the 2012 Rio+20 document *The Future We Want*, or the documents that have come out of the United Nations Climate Change Conferences from COP-1 in 1995 to the most recent COP-21 in 2015, we may view these very crudely as a series of failed attempts to resolve the trauma of the eco by making the world become one in order to correspond to the earth that is exactly one. The thinking behind these declarations seems to be that as long as each individual subject or each individual country is focused merely on its own particular interests and needs, political efforts to address ecological issues remain limited by the individual world-views of these subjects or countries. The solution to the fragmentation of politics into individual world-views is thus the constitution of one all-encompassing collective subject that is able to grasp the totality of the world. This would be a subject that is able to perceive and translate the earth into one world. To find out what is this subject we only need to ask the following question vis-à-vis these declarations: Who speaks? And the simple answer is: *We* speak!

Among all the international and global summits whose purpose is constituting a collective 'We' that is capable of recognizing, addressing and solving global ecological issues, the COP-15 in Copenhagen 2009 stands out as the most blatant failure. In the period leading up to the summit, environmental sceptics questioning the causal links between human activity on the one hand and global warming and climate change on the other had finally been forced out of mainstream political discourse. Hence a major obstacle in the road towards global responsibility and action seemed to have been cleared. The COP-15 conference in Copenhagen 2009 was supposed to be the moment when the world would finally come together and act in unison to prevent ecological disaster. In an advertising campaign, the city was even branded as 'Hopenhagen', and the stated mission of the campaign was the following:

> Hopenhagen is a movement, a moment and a chance at a new beginning. The hope that in Copenhagen this December – during the United

Nations Climate Change Conference – we can build a better future for our planet and a more sustainable way of life. It is the hope that we can create a global community that will lead our leaders into making the right decisions. The promise that by solving our environmental crisis, we can solve our economic crisis at the same time.[16]

As we know now, these expectations were far from redeemed and the conference ended in a kind of Mexican stand-off (ironically the next conference, COP-16, was set to take place in Mexico), with each of the world powers refusing to commit themselves to political action unless the others would do something first. Skipping through the intricacies of realpolitik, the fundamental reason for this collective impotence was that each of the parties had already committed themselves unconditionally to the premises of global capitalism. And within this framework, only solutions leaving the fundamental conditions of capitalism intact – the imperative of growth, free trade, free movement of capital and so on – are imaginable. It is symptomatic that so much faith is put in science coming up with smart solutions to environmental problems. Basically, 'smart' here means something that will solve our problems without requiring us to change the basic coordinates of consumer capitalism. In this sense, the limitations of solving the environmental crisis were perhaps already inscribed in the hopes invested in the conference. The last sentence of the above mission of the Hopenhagen campaign reads: 'The promise that by solving our environmental crisis, we can solve our economic crisis at the same time.' This elegantly rules out the possibility that the economic crisis *is* precisely the solution to the environmental crisis.

This point may be illustrated by the well-known press photo (Figure 2.2) showing Obama along with key European leaders during the COP-15 negotiations in Copenhagen. The COP-15 conference may be seen as testimony to the psychoanalytic insight expressed through the Lacanian slogan 'The big Other does not exist.' The failure of the conference was a demonstration that 'We' does (*sic*) not exist. The collective subject supposed to act not only on behalf of all the people of the world but even on behalf of the totality of the earth as such is ultimately a phantasmatic image. Of course, fantasies may be highly functional as individual subjects incorporate them and act as if they are real. On the other hand, fantasies may also turn out to be highly fragile when we start insisting that they manifest themselves positively in the world. The problem of the COP-15 as well as similar summits is that, when world leaders come together, the vague phantasmatic image of a collective We is translated into a collection of concrete individual subjects. However, this reveals the embarrassing

Figure 2.2 COP-15 meeting

truth that no one is this We. The hope invested in Hopenhagen was that We (the big Other) would emerge at the event and start taking care of all the problems that we (the conglomeration of individual subjects) have been unable to solve.

This point may be illustrated by the above photograph showing Obama along with key European leaders during the COP-15 negotiations in Copenhagen. While there is certainly a lot of powerful people gathered in the room, there is at the same time also something strangely mundane about the image. It is as if there is someone missing from the photo (and this missing someone is not the then Chinese premier Wen Jiabao, who was conspicuously absent from the whole summit). The philosophical point here is of course that the big Other is missing from the picture. The image functions to bring us 'too close to the sublime objects of power', whereby they become stripped of the phantasmatic properties that we otherwise tend to project onto them as they appear in normal settings staged behind a podium, exiting an airplane or in an arranged group photo. The people we see in the picture are not Obama, Sarkozy, Merkel, Brown and Rasmussen but rather just Barack, Nicola, Angela, Gordon and

Lars. The image gives anything but the impression that what we are looking at is the global political We in resolved negotiation. Instead, what we see is a bunch of individual people scratching their noses, checking their emails or just trying to pretend that Lars Løkke Rasmussen is saying something important. In this sense, the image is a very accurate account of the COP-15, and one might be tempted to supply it with the title: 'COP-15 when the We had left the building.'

It is interesting to read the resolution from the 2012 Rio+20 Earth Summit in light of the breakdown in Copenhagen. As noted, the text is entitled *The Future We Want*. The first chapter of the text is entitled: 'Our Common Vision' and the first paragraph of the chapter reads:

> 1. We, the Heads of State and Government and high-level representatives, having met at Rio de Janeiro, Brazil, from 20 to 22 June 2012, with the full participation of civil society, renew our commitment to sustainable development and to ensuring the promotion of an economically, socially and environmentally sustainable future for our planet and for present and future generations.[17]

In total, the text consists of 283 such paragraphs. With only a few exceptions, all of these paragraphs begin with the subject predicate 'We', and in the text comprising in total 53 pages this predicate appears 627 times. The text of course invites the immediate interpretation that here speaks the global collective subject finally emerging with strong resolve to save the planet from global climate change and other catastrophes. But taking a step back from the content of the text, there is something almost neurotic about the constant and insistent repetition of the word 'We'. Rather than a resolution, the text seems to give the impression of a conjuration or even a prayer: If we keep talking as if this 'We' exists, then finally it must emerge. The text even seems to serve the purpose of persuading the authors themselves that what is in the text is actually true.

The Continuation of Protestantism by Other Means

The preceding analysis has explored how ecology has emerged as a certain way of thinking and talking about nature. By the appropriation of Žižek's analytical framework, we have been exploring ecology as a particular symbolic order aiming to symbolize the eco. A fundamental axiom in Žižek is that any symbolic order is constituted by a 'lack' of symbolization, a point where symbolization is impossible.

This lack is not just a mark of the insufficiency of the symbolization. It functions as the very structuring principle of the symbolic order. Žižek states that 'the symbolic field is in itself always already barred, crippled, porous, structured around some extimate kernel, some impossibility'.[18] He even goes on to state that this impossibility is the very condition of possibility for symbolization and for the constitution of social reality: 'Reality itself is nothing but an embodiment of a certain blockage in the process of symbolization. For reality to exist, something must be left unspoken.' '[T]here is "reality" only in so far as there is an ontological gap, a crack, at its very heart – that is, a traumatic excess, a foreign body that cannot be integrated into it.'[19]

These definitions may serve to summarize the argument of the preceding analysis. The most significant point in this chapter is the demonstration that ecology is unable to integrate Man into the account of Nature. Paraphrasing Žižek, human subjectivity constitutes the 'impossibility', the 'extimate kernel', around which the symbolic field of ecology is structured. The subject is 'left unspoken' by the ecological account of the world, and Man thus becomes a 'traumatic excess, a foreign body that cannot be integrated into' the ecological account of Nature. At the same time, this impossibility of integrating the human subject into the symbolic order of ecology is also the very possibility of ecology. When natural phenomena do not conform to preconceived notions about nature, balance and harmony, they may be explained away with reference to human interventions in the natural order of things. We can find a similar principle at work in neo-classical economics adhering to free-market ideologies, where occasional market failures are explained away with reference to disturbing interventions by non-market participants such as the state.

As ecology evolves from merely a scientific discipline into also being a significant component in contemporary politics, the inability to account for the human subject as anything but an intrusive and destructive agent, the polluting animal, makes ecology inefficient as an ideological vehicle for the kinds of political mobilization required to reverse or even just curb the current trends of ecological destruction that may eventually lead to global catastrophe. Ecology does not offer the individual subject very many meaningful points of identification. It is very efficient in pointing out the shortcomings of human nature and human societies in terms of their effects on natural ecosystems, but it is much less efficient in pointing out positive human potential. As already suggested in the quote from Žižek introducing this chapter, ecology is ultimately a somewhat misanthropic ideology, if we follow its logical implications to the end. In order to live in complete concurrence with ecology, humans should either learn to be more like animals, that is to de-learn that which makes us human in the first place, or simply just disappear from the planet altogether.

Now of course this is a very crude and extreme reading of ecology. It is therefore important to note that the purpose of the reading is not to discard the meaningfulness of all the practical things that we can do in order to have a more ecological lifestyle such as recycle, buy organic products, limit our use of fossil fuels and so on. On the contrary, the point of the reading is to understand why ecology has been and still is a relatively ineffective ideology in terms of changing people's lifestyles as well as their economic and political priorities in ways that have a significant impact on global production and consumption patterns. And one answer would be that most people find it difficult to identify fully with an ideology that ultimately denounces most of their immediate desires. When ecology is appropriated not only as a scientific method for the study of ecosystems but also as a theory for understanding human nature and the relation between humans and nature, it invariably tends to produce guilt. As one becomes aware of the detrimental impact of human activity on the natural environment, it is difficult not to become overwhelmed by guilt on behalf of the entire species. The only kind of consolation is pointing to other individuals and other groups of human beings that are more un-ecological than oneself.

The preceding genealogical analysis has shown how the concept of the balance of nature is gradually decoupled from the religious superstructure out of which it originated as ecology evolved into an independent scientific discipline. Still, there are significant theological issues that seem to be hibernating within the ideology of ecology. We have already looked into the ecological version of the problem of theodicy. Guilt is another theological issue in ecology. We may recognize the Christian notion of original sin in the account of the human subject implicitly found in ecology. If it is part of the ontology of humans to transcend and disturb the inherent balance of nature, then the mere fact of being born a human being is a disturbance to the balance of nature. Hence all humans ultimately carry the burden of ecological guilt. The only way of acting on this guilt is by limiting your impact on nature. The ecological duty of the human subject is exercising damage control on the catastrophic event of one's own birth. The metaphor of carbon footprint is illustrative of this point. The lighter you tread on the face of the earth, the better, and the perfect ecological human would be someone capable of hovering just above the ground, leaving no trace of his or her own existence in the world. We can also note how political goals for environmental issues are typically formulated in strictly negative terms. The ambition of these goals is not to do something that is good in itself but rather to limit the bad consequences of our being on the planet. The 1997 Kyoto Protocol provides perhaps the best illustration for this point as it commits a number of industrialized countries to reduce their emissions of greenhouse gases by an average of 5.2 per cent compared with the benchmark year of 1990. In political and economic terms, this may be regarded as an ambitious target. However, in philosophical or even theological terms there seems to be something very despondent about this way of formulating the aspirations of mankind. Rather than coming together around some grand positive ideal, we should merely just aim to reduce the impact of our own failed being on the planet. The fact that Denmark, one of the most thoroughly Protestant countries in the world, is at the forefront of implementing these global policies may be taken as an indication of the Protestant heritage in this kind of ecological policy making.

So identifying with the ideology of ecology invariably means that you accept the burden of shame and guilt inherent in the fact of being a human in a natural world. While this notion of ecological guilt is similar to the Christian idea of original sin, there is a crucial difference between ecology and Christianity at this point. Together with the acceptance of guilt, Christianity also offers the possibility of divine love and forgiveness. Even though you are just an imperfect human being, God still loves you and He is willing to forgive you. Ecology does not open up this option. There is no authority with the mandate to offer any kind of grace for the polluting animal. Mother Nature may be hurt and disappointed with you as a human being but there is no heavenly Father to punish or forgive or otherwise bring justice and love into the world of ecology.

Part II

The Nomy of Eco

3

How *is* the Economy?

Global prospects have improved again but the road to recovery in the advanced economies will remain bumpy. World output growth is forecast to reach 3¼ percent in 2013 and 4 percent in 2014. In the major advanced economies, activity is expected to gradually accelerate, following a weak start to 2013, with the United States in the lead. In emerging markets and developing economies, activity has already picked up steam.[1]

This passage comes from the introduction to the 2013 *IMF World Economic Outlook* carrying the subtitle: *Hopes, Realities, and Risks*. The particular content of the passage or even the report in its entirety is not so important in the present context. In principle, we could have picked another passage from any number of similar reports. What is interesting in the context of eco-analysis is the way that the text speaks about the economy. As we have seen, economy and ecology are very closely related in terms of their etymological roots. They only differ in terms of their respective suffix that may be taken to mean either law (-nomy, *nomos*) or discourse and logic (-logy, *logos*). However, there is a very remarkable difference in the way that we apply the two terms. When we speak of ecology, we do not treat it as an entity in itself. We speak of ecology as the study of particular natural phenomena or even as the study of ecosystems. Or we speak of the ecology *of* something, for instance the ecology of plants, the ecology of a lake or even the ecology of the city. Ecology is, in other words, used in conjunction with an object, a place or a phenomenon to indicate a certain way of relating to that object, place or phenomenon.

This is different from the way we use the word 'economy'. Of course, we can also speak of the economy *of* something, such as the economy of farming, the economy of fishing or the economy of cities. But, contrary to ecology, the word 'economy' may stand alone as an entity in itself. This is what happens, when we use economy in sentences such as: 'China's economy is expected to expand at 7¾ per cent this year'; 'Barack Obama is not good for the economy'; or 'the economy does not allow us to lower taxes at the moment'. In this way, we think of the economy as an object or even a subject in itself. The passage from the IMF report serves to illustrate this. Although it speaks of economies in the plural, these are still treated as entities to which different qualities can be attached. In comparison, it would not make sense to speak of the 'ecology' as such. Instead, we would need to talk about the ecosystem, the biosphere, the natural environment or some other entity with an independent positive existence.

This difference in our use of the words 'economy' and 'ecology' may be seen as an illustration of their respective success and failure in shaping the ideology of society. As ideology, economic thinking has succeeded to a point where we no longer distinguish between economy as a symbolic order and the object that is being symbolized. We accept the economy as an independent entity and agent out there in the world. This means that when we act in the world, we take into account the possible effects of our actions on the economy. Furthermore, we also experience the economy as acting upon us and constantly having real effects on our being. In philosophical terms, the gap between reality and the real appears to have been closed in the case of the economy. This is what makes the economy so effective as an ideology shaping our ways of thinking, speaking, acting and desiring. In contrast, we rarely if ever have an experience that the 'ecology' (the quotation marks indicate that the word barely makes sense) intervenes in our life. Nature may of course intervene in our lives, for instance when it rains, when we are bitten by a wasp or when our house is destroyed by a hurricane. However, we do not readily think that such events are the expression of the 'ecology'. At best, we can have an abstract idea that a higher frequency of hurricanes may be the result of changing conditions in the global climate system caused by CO_2 emissions from the burning of fossil fuels.

Another way of conceptualizing the difference between ecology and economy is to suggest that ecology is a way of speaking about something else, whereas the economy has the capacity of speaking for itself. Ecology does not have its own language, whereas the economy is able to express itself through the language of money. This means that changes to the state of the economy are readily expressed in terms of money. Prices of commodities go up or down. Interest rates go up or down. And the productive output of nations as measured in gross domestic product (GDP) goes up or down. All of these changes may be recorded through the same medium: money. The IMF quote exemplifies this very precisely. It is a very simple and immediate account of the state of the economy. However, when the state of an ecosystem changes, this does not find its expression in any unitary medium. Instead, we find a multitude of media for measuring the state of various ecosystems. The concentration of CO_2 in the atmosphere is measured by parts per million (ppm). Temperature levels are measured by degrees that may be compiled to express global averages. Biodiversity is measured by the amount of animal and plant species that have not yet been exterminated. And so on. As an academic discipline, much of ecology is engaged in performing and developing these kinds of measurements. This obviously involves a complicated process of translation, where the state of ecosystems is given a symbolic expression. Again, ecology is this process of symbolization, whereas the economy is itself an object of symbolization. This is why the economy is so much more effective in establishing itself as an entity to be taken into consideration in our private and political lives.

Consider here the difference between the following two propositions: (1) 'I don't believe in ecology'; (2) 'I don't believe in the economy.' We can image the first proposition being uttered by a politician who is sceptical of predominant theories of global warming or maybe a farmer who is sceptical of the benefits of ecological farming. While we may disagree with someone making such a proposition, at least we can accept it as part of a meaningful conversation. However, the second proposition is on the verge of being meaningless. It almost amounts to saying 'I don't believe in gravity' or 'I don't believe in the sun.' The economy is a fact that will have an impact on our lives, whether we believe in it or not.

In this chapter, we shall shift the focus of our eco-analysis from ecology to economy. The point of departure for the analysis is the problem of growth, which stands at the heart of the ideology of contemporary economy. It seems that, because of the way our economies are organized today, they need to perform perpetual growth in order to function properly. Unless GDP grows by a certain percentage every year, national economies become dysfunctional, resulting in rising levels of unemployment and of debt. The ideological success of economic thinking today means that, within a wide range in the political spectrum, there is a consensus that the top priority of the political management of a country is to perform and maintain certain levels of economic growth. Contemporary economics as well as contemporary politics seems to be structured around an imperative of growth.

The problem of growth is also interesting in the context of eco-analysis because it constitutes a nodal point in the relation between economy and ecology. Growth is ultimately a natural phenomenon that we connect with the life and death of plants and animals. To some extent, we might even say that growth is the same as life. A fundamental characteristic of living organisms is their capacity for growth. And, once an organism stops growing, it is already in the process of dying. However, when the notion of growth is carried over into the domain of economy, it comes to mean something completely different while at the same time retaining some of the same connotations. Within politics, the problem of economic growth often marks the limit of ecology. Few politicians are against ecological politics per se, such as natural preservation, sustainable production, anti-pollution measures and so on. However, support for such policies often withers away when their implementation is perceived to compromise economic growth. Instead, the bulk of political initiatives addressing ecological issues seeks to combine ecological measures with economic profitability. We have already touched upon this combination of ecology and economy, which constitutes the third option in the matrix built around the Lenin joke at the beginning of the book.

The purpose of the following eco-analysis is to uncover the ontological constitution of the concept of growth in contemporary economic thinking. The above-mentioned ideological success of economy is partly due to the fact that it is very capable of presenting itself as a complete and consistent symbolic order. In this respect, it differs significantly from ecology, which appears as a much more fragile and fragmented symbolic order. Yet the idea here is to wrench open the gap between economy and the eco, between reality and the real, thus opening new potentials that seem to be veiled in the prevalent way of thinking about economy and economic growth in particular.

We can use the IMF quote from the beginning of this section as a point of departure for the analysis. But instead of moving forward from the quote and into the economic analyses of the report, we can move backwards by posing the simple question: If this is the answer, what is the question? And the answer to this question is equally simple. The IMF report provides an 'economic outlook' on the world, so the implicit question preceding the quote can be posed as 'How is the economy?' Now, in the following, we shall be working from this question but allowing ourselves the liberty to tweak its meaning slightly. The question 'How is the economy?' may immediately be taken to mean: How is the economy *doing*? This is the same kind of question that we might ask a friend, when we meet him: 'How are you doing?' And he might proceed by listing some of the things that are currently most important to his well-being, much in the same way that the IMF lists those features of the world economy that seems most pertinent to its overall well-being.

However, if we submit the question to a Heideggerian reading, it contains another possible interpretation. Of course, Heidegger himself did not care much for the economy, let alone a philosophical exploration of economy. Nevertheless, his way of asking questions may still be appropriated precisely for this kind of exploration. When we ask questions such as 'What is man?', 'What is money?', 'What is war?', 'What is the economy?', our thinking is immediately directed to what Heidegger refers to as the ontic domain of beings. Asking in this way, we tend to take for granted the meaning of 'is' and we are merely concerned with the definition of entities in relation to other entities. What is man as opposed to an animal? What is money as opposed to a commodity? What is war as opposed to peace? What is the economy as opposed to politics? In order to think about the very Being of things, to think about existence, we need to pose our questions in a different fashion. Rather than glossing over the meaning of 'is', this should be the very focus of the question. Now Heidegger himself was concerned with Being itself and his inquiries into the Being of specific beings were all merely occasions for the exploration of Being itself. Along these lines, he proposes that we ask the question: How *is* Being?[2] The point of the italics is to emphasize that it is the very meaning of 'is' in the question which is the real enigma to be explored. Translating this way of posing philosophical questions in the context of our eco-analysis of economy, the original IMF question is altered into the following: How *is* the economy? Now all of a sudden we are not asking about the immediate state of the economy and its well-being. Instead, we are asking about the very ontological constitution of economy. This is what we are going to explore in the following.

Where Does Value Come From?

> Indessen begannen Härte und Geruch des Eichenholzes vernehmlicher von der Langsamkeit in Stete zu sprechen, mit denen der Baum wächst. Die Eiche selber sprach, daß in solchem Wachstum allein gegründet wird, was dauert und fruchtet: daß wachsen heißt: der Weite des Himmels sich öffnen und zugleich in das Dunkel der Erde wurzeln;

daß alles Gediegene nur gedeiht, wenn der Mensch gleich recht beides ist: bereit dem Anspruch des höchsten Himmels und aufgehoben im Schutz der tragenden Erde.[3,4]

At the heart of the most prevalent contemporary accounts of economy lies the notion of growth. How can individual companies grow their production, turnover and profits? How can nations grow their collective wealth by growing their productive output as measured in GDP? However, neither of these questions is the concern of eco-analysis. Instead of moving forward from these questions of economic growth, we shall be moving backwards in order to uncover the notion of the economy that lies inherent in this way of posing the problem of growth.

Our investigation of economy takes an unlikely beginning with Heidegger. In the above passage from his brief poetic text on *Der Feldweg*, Heidegger speaks about the growth of an oak tree. Or perhaps it is more appropriate to say that, through the text, Heidegger allows the oak itself to speak about growth. The oak has been cut down. The father brings an allotted cord of the wood to his workshop and his boys carve out little ships from its bark. It is in this careful processing of the wood that the time which has gone into the growth of the oak emanates from very material being of the wood. If we go along with Heidegger and the oak, we are provided with the following definition: 'growing means: to open oneself to the expanse of the heavens as one takes root in the darkness of the earth'. In the context of Heidegger's philosophy, this statement is not merely an account of the growth of an oak tree but serves as a metaphor for the Being of Man himself, who is simultaneously grounded in the material world of the earth and opened towards the heaven of ideas and thinking.

When the IMF says that in the third quarter of 2012 the German economy grew by 0.90 per cent, they are obviously not referring to the same thing that Heidegger's oak is speaking about. Economic growth is typically measured in GDP. A standard definition from the IMF reads: 'GDP refers to the total market value of all goods and services that are produced within a country per year. It is an important indicator of the economic strength of a country.' In other words, when GDP increases from year to year, we can speak of economic growth. We are here back to the conventional problem of economic growth stated above.

This juxtaposition between the oak's and the IMF's definition of growth may serve as a point of departure for the very opening of the problem of growth. Before we plunge into an investigation of the problem of growth, we need to consider properly how to pose this problem as a question. Perhaps the most obvious way of posing the problem of growth is by asking: What is growth? This question might foster an inquiry activating sub-questions such as: How is economic growth measured? What does economic growth measure? What doesn't economic growth measure? What is the relation between economic growth and other forms of growth, such as ecological growth, spiritual growth, growth in welfare and well-being? What are the limits to economic growth? And what are the prospects for capitalism beyond growth? As much as all of these questions are interesting and highly pertinent to contemporary society, eco-analysis takes another approach to the issue of growth that does not involve a direct engagement with this kind of question. And again, instead of moving forward *into* solving the problem of growth, we move *backwards* by thinking about the very nature of the problem of growth. Our first question thus becomes: What is the problem of growth?

If it were not for the fact that we live in a time that is hysterically obsessed with the notion of economic growth, most people would probably not even think about growth as an economic concept. Originally, growth is an inherent property of life itself: trees grow, flowers grow and sheep, sharks and babies grow. This is perhaps the understanding that Heidegger is aiming at as he allows the oak to speak to us about growth. But even when humans speak about growth, the concept seems to belong first and foremost in the sphere of ecology as we think and talk about the way that the growth of natural organisms is tied into a larger natural cosmology. It is only when the result of natural growth becomes objects of human consumption and exchange that the notion of growth moves into the sphere of the economy. Once we start thinking about the corn growing in the field as a source of food or income, or the little calf growing into a cow that may be eaten or sold as beef, growth is inscribed into a symbolic order of economy. At this point, the growth of natural organisms is conceived in terms of an increase in the amount of commodities available for consumption and exchange within an economic entity. This point marks the parallactic shift between ecology and economy.

Today we have become accustomed to talking about economic growth in terms of GDP. Inherent in our contemporary measure of GDP lies a number of fundamental assumptions about value, price and growth. For instance, measurements of GDP take the market price as the true representation of the value of a commodity. Measurements of GDP also only include those commodities and services that are made available for exchange and consumption in the market. And comparisons of annual GDP do not account for any decline in the natural resources available for the economy. Once we start poking into the philosophical intricacies of the relation between value and price, we also see that the notion of economic growth becomes highly elusive. Measuring economic growth presupposes some kind of pricing mechanism that allows us to compare the value of the different commodities and services, as well as the value of productive output of an economic entity in year one with the productive output in year two.

The contemporary method of measuring the value of an economy in terms of GDP is merely one way of rendering the eco into an economic order. In the following, we shall be unfolding three different theories of economics, each representing a distinct answer to the question 'Where does value come from?' and thus also three distinct notions of economic growth. The presentation of these three economic paradigms of growth is structured along the lines of Žižek's threefold ontological distinction between the real, the symbolic and the imaginary. The purpose of this philosophical mapping of different theories of economic value is not to derive a true or correct notion of growth. On the contrary, the ambition is to demonstrate the elusive nature of economic growth while at the same time constructing a theoretical framework that allows us to analyse contemporary problems related to growth. As mentioned previously, we are simply shifting our eco-analysis from the domain of ecology to the domain of economy. The purpose is to flesh out three ideal-typical forms of thinking about value and economic growth. This means that the readings are deliberately tendential as we are going to emphasize that particular dimension of a dimension that fits with our overall distinction between the real, the symbolic and the imaginary.

First, we shall be looking into the school of physiocrat economics of value found in François Quesnay and Anne Robert Jacques Turgot with the purpose of exemplifying a substance theory of value. This theory aims to found value in the domain of the real as it views the produce of the land as the basis of all economic activity. Second, we shall be looking into the paradigm of classical economy found in Adam Smith and David Ricardo. Our reading of these classical economists goes beyond their internal differences and inconsistencies in order to present a labour theory of value. Since labour is the cultivation and processing of natural material, we may understand this form of thinking as a foundation of the economy in the domain of the symbolic. Finally, we shall be analysing the neo-classical paradigm of economy, which is at the heart of the contemporary notion of GDP as a measure of growth. The primary reference point is the production function originally formulated by Robert Solow and Trevor Winchester Swan and subsequently advanced by Paul M. Romer, among others. In neo-classical economics, we find a market theory of value, where the problem of value seems to have been solved through the notion of the efficient market. The market figures as the big Other capable of determining the true value of commodities and services. Hence the pivotal point of the economy has shifted onto the domain of the imaginary, where the difference between the economy and the market is collapsed.

The Substance of Value

In 1758, François Quesnay published the *Tableau Economique*, shown in Figure 3.1.[5] The revenue of 600 m livres at the top of the table represents the net produce of a farmer. Out of an annual agricultural gross production worth a total of 1,500 m livres, 600 m livres-worth of produce is immediately used for the maintenance of the farmer as well as his family, workers and livestock. Another 300 m livres is spent on non-farming goods purchased from the class of artisans and merchants. This amount figures at the top right-hand side of the table. These deductions leave a net produce of 600 m livres, which is paid in rent to the class of landowners. The table now tracks how this revenue flows throughout the economy as the landowners use the revenue for purchases from other classes. Quesnay's economic table is arguably one of the earliest attempts to analytically describe the economy of a state as a system of interacting parts, and he should thus be regarded as one of the founders of political economy.

Quesnay belongs to the school of economic thinkers known as the physiocrats. What characterizes this school of thinkers is their emphasis on the land as the ultimate source of all value. This idea is succinctly expressed by one of Quesnay's fellow physiocrats, Anne Robert Jacques Turgot, writing the following:

> §7. The husbandman is the only one whose industry produces more than the wages of his labour. He, therefore, is the only source of all Wealth.
>
> The situation of the husbandman [compared to the mere workman – OB] is materially different. The soil, independent of any other man, or of any agreement, pays him immediately the price of his toil. Nature does not bargain with him, or compel him to content himself with what is absolutely necessary. What she grants is neither limited to his

DÉPENSES PRODUCTIVES.	DÉPENSES DU REVENU, l'Impôt prélevé, se partagent aux Dépenses productives & aux Dépenses stériles.	DÉPENSES STÉRILES.
Avances annuelles.	*Revenu.*	*Avances annuelles.*
₶	₶	₶
600 produisent	 600	300
Productions.		Ouvrages, &c.
₶	₶	₶
300 reproduisent net	 300	300
150 reproduisent net	 150	150
75 reproduisent net	75	75
37..10 reproduisent net	37..10	37..10
18..15 reproduisent net	18..15	18..15
9....7....6 reproduisent net..	9....7....6	9....7....6
4..13....9 reproduisent net..	4..13....9	4..13....9
2....6..10 reproduisent net..	2....6..10	2....6.10
1....3....5 reproduisent net..	1....3....5	1....3....5
0..11....8 reproduisent net..	0..11....8	0..11....8
0....5..10 reproduisent net..	0....5..10	0....5..10
0....2..11 reproduisent net..	0....2..11	0....2..11
0....1....5 reproduisent net..	0....1....5	0....1....5

Figure 3.1   Tableau Economique

wants, nor to a conditional valuation of the price of his day's work. It is a physical consequence of the fertility of the soil, and of justice, rather than of the difficulty of the means, which he has employed to render the soil fruitful. As soon as the labour of the husbandman produces more than sufficient for his necessities, he can, with the excess which nature affords him of pure freewill beyond the wages of his toil, purchase the labour of other members of society. The latter, in selling to him, only procures a livelihood; but the husbandman, besides his subsistence, collects an independent wealth at his disposal, which he has not purchased, but which he can sell. He is, therefore, the only source of all those riches which, by their circulation, animates the labours of society: because he is the only one whose labour produces more than the wages of his toil.[6]

The revenue of 600 livres in Quesnay's table represents the surplus produced by the class of farmers over and beyond the produce required for the reproduction of this class and its livestock, as well as the purchase of necessary goods from the class of artisans and merchants. It is the production of this surplus, which enables the economy to allocate labour in non-farming forms of production, to develop more advanced forms of division of labour, and to develop forms of exchange beyond the simple household. In the words of Turgot, the farmer is 'the first mover of the whole machine of society'.[7] Both Quesnay and Turgot distinguish between three different classes in society: the agricultural class; the class of artisans and merchants; and the class of landowners. Quesnay, who also had a background in medicine, refers to the three classes as the 'productive', the 'sterile' and the 'proprietary' class. Along similar lines, Turgot speaks of the 'productive', the 'stipendary' and the 'disposable' class.

What strikes the contemporary eye is that, in the view of the physiocrats, even artisans and merchants are not considered to be productive. While labourers of this stipendary class do indeed perform vital functions in society, their labour does not produce any surplus value over and beyond what is needed for their subsistence:

> The two classes of cultivators and artificers, resemble each other in many respects, and particularly that those who compose them do not possess any revenue, and both equally subsist on the wages which are paid them out of the productions of the earth. /.../ But there is this difference between the two species of labour; that the work of the cultivator produces not only his own wages, but also that revenue which serves to pay all the different classes of artificers, and other stipendiaries their salaries: whereas the artificers receive simply their

salary, that is to say, their part of the productions of the earth, in exchange for their labour, and which does not produce any increase.[8]

The argument here is that the artificer is paid a salary, which is the equivalent of the value of his production, while the cultivator (the farmer) is subject to a form of exploitation in so far as the fruits of his labour exceed the value of the wage he receives. The surplus value created by the farmer is of course appropriated by the proprietor of the disposable class, whose only economic function is to distribute this revenue in society through consumption. Since the land is the ultimate origin of value and the cultivators the only class capable of producing surplus value, this is also where we find the source of possible growth in the economy.

The main trait in physiocrat economics is a substance theory of value. Value is ultimately derived from the organic substance produced by the land. To capture the nature of this substance of value, Quesnay also uses the concept of *blé*, which translates into 'corn' or 'wheat'.[9] This also means that the notion of economic growth held by the physiocrats is closely connected with the notion of growth found in the discourse of ecology or even just in our immediate common-sense understanding, where growth is a feature of the life of natural organisms. For the physiocrats, economic growth of a nation is achieved by the expansion of the areas of land cultivated for the purpose of farming, forestation or other kinds of natural production or by the improved cultivation of existing areas of land. The purpose of such efforts is the increase and optimization of the growth of natural organisms that may ultimately become the objects of human consumption. In other words, the economy of a nation may grow as the natural growth of corn, grapes, potatoes, sheep, cows, trees and so on within that nation is increased.

At the same time, the physiocrat notion of growth does not coincide entirely with the purely ecological notion of growth. Economic growth does not comprise the growth of all natural organisms but only those organisms that become the direct or indirect objects of human consumption. The growth of wheat is part of a growing economy, while the growth of weeds is not. Even though the physiocrats put nature at the centre of the economy, their thinking still rests on a fundamental distinction between nature and culture, the wild and the domestic. The farmer is referred to as the cultivator, and only those parts of nature which are subject to human cultivation are also considered part of the economy. All economic value may ultimately derive from a natural substance but not all natural substances are necessarily economically valuable. As we have already seen in the first part of this book, ecology takes a wholly decentralized view of nature as an entity with intrinsic value and purpose and with inherent capacities for sustainability and balance. In ecology, the growth of weeds is as important as the growth of wheat. Both are part of the same system.

Price and Value

Quesnay is not only regarded as one of the founders of political economy but also a forerunner in the field of quantitative macroeconomics, as he aims to describe the economy of a nation in terms of monetary figures. This is very obvious in the economic table, where the flow of value throughout the economy is mapped out in terms of the amount of money representing this value. Value may ultimately have the nature of the substance *blé* but the only way of describing it is through the symbolic price of money. Invoking again the perspective of eco-analysis, we can recognize the description of value in terms of price as a symbolization of the real. Economy emerges as the eco is subjected to the distinction between value and price. In fact, this is true not only in the specific case of the physiocrats but perhaps for all forms of economic thinking that comes to found an economic system. Let us take a minor excursion from our analysis of the physiocrats in order to explore the distinction between value and price from an eco-analytical perspective.

It is tempting to think that economy is merely a system for the pricing of value that is already an inherent property in the reality of the eco. Corn and cows have intrinsic value as the source of human nutrition in the form of flour, bread, milk and beef. Economy is the eco-naming of this value in terms of price. However, this idea of the pricing mechanism is too simple. While it is true that value belongs in the order of the real, this is not the same as saying that value is simply there in a way that precedes the symbolic operation of pricing. The distinction between price and value thus corresponds to Žižek's distinction between symbolic and real.[10] On the one hand, a price seems to be merely a symbolic expression of the real value of an entity or event. When a loaf of bread sells for 10 kroner, this seems to be an expression of the real inherent value of the bread. On the other hand, we can also view value as nothing but the subsequent projection of the symbolic price onto the domain of the real. The bread has value simply because it sells for 10 kroner. The philosophical answer to this apparent contradiction is of course to say that value is both one and the other. Value is real. It is both the starting point of symbolization and it is also a retroactive projection emerging from the process of symbolization. We may again utilize Žižek's account of the parallax real in order to further unpack this intricate relation between value and price:

> The parallax Real is thus opposed to the standard (Lacanian) notion of the Real as that which 'always returns to its place' – as that which remains the same in all possible (symbolic) universes: the parallax Real is, rather, that which accounts for the very multiplicity of appearances of the same underlying Real – it is not the hard core which persists as the Same, but the hard bone of contention which pulverizes the sameness into the multitude of appearances. In a first move, the Real is the impossible hard core which we cannot confront directly, but only through the lenses of a multitude of symbolic fictions, virtual formations. In a second move, this very hard core is purely virtual, actually non-existent, an X which can be reconstructed only retroactively, from the multitude of symbolic formations which are 'all that there actually is'.[11]

The idea is that value should not be considered as something 'which remains the same in all possible symbolic universes' or a 'hard core which persists' beyond any kind of procedure of pricing. Instead, we can describe the interrelation between the real of value and the symbolic operation of pricing through the following paraphrase: 'In a first move, value is the impossible hard core which we cannot confront directly but only through the lenses of a multitude of price formations. In a second move, the very hard core of value is purely virtual, actually, non-existent, an X which can be reconstructed only retroactively, from the multitude of price formations.'

What is the value of a bushel of corn? On the one hand, we can only speak of the economic value of an object through some notion of price. It is only through some kind of pricing that we are able to compare the value of different objects. On the other hand, no process of pricing is capable of fully capturing the value of the object. The reason why exchange takes place in the market is precisely because of the elusive nature of value. One party agrees to sell a bushel of corn for 150 roubles because to him the corn is worth less than 150 roubles. If the corn was worth more than 150 roubles to him, he would not sell it. In turn, another party agrees to buy a bushel of corn for 150 roubles because to him the corn is worth more than 150. If this were not the case, he would of course just keep his money and not buy the corn. For the exchange to take place, the seller and the buyer need to agree on a certain price but at the same time they need to disagree about the real value of the corn. In the same way that the subject in Žižek's thinking is a symptom of the traumatic split between the real and the symbolic representation of the subject, so the market is a symptom of the traumatic split between the real value and its symbolic representation through price.

As we return to the specific analysis of the physiocrats, we can see how the traumatic split between value and price stands out as an open wound within their description of the economy. Both Quesnay and Turgot are writing at a time in history when the French economy is subject to an extreme degree of government control and regulation, with prices being determined largely by the state rather than by the kind of free-market mechanisms that we know today. On the one hand, Quesnay's table is an attempt to describe the way that value and money actually flow through the economy. On the other hand, the physiocrat theories of Quesnay and Turgot have a strong normative element as they believe that the actual flow of value and money in their contemporary French economy is both economically inexpedient and morally unjust. In brief, their arguments may be summarized as a critique that the current pricing of agricultural produce is wrong.

The origin of value and the source of surplus produce enabling the economic growth of the nation are the land, and therefore a larger share of this value should flow back into the agricultural class rather than being appropriated by the proprietary class and wasted away on an 'excess of luxury in the way of ornamentation'.[12] This redistribution of the resources of the nation constitutes an alignment of the price of agricultural commodities with their true value. In concrete terms, this amounts to a call for tax reform shifting the burden of taxation from the cultivators to the owners of the land, as well as a call for free trade to allow prices to be determined by the market rather than the state. Quesnay ends his work on the economic table with the following manifesto, which seems to already hold the seeds of the imperative of growth permeating contemporary economics and politics:

> The territory of France could produce as much and even much more. /.../ We are speaking of an opulent Nation which possesses a territory and advances which yield annually, and without wasting away, 1 billion 50 millions of net product; but all these riches kept up successively by this annual product may be destroyed or lose their value, in the decadence of an agricultural Nation, by the mere wasting away of the productive advances, which may make great headway in a short time as a result of eight principal causes:
>
> 1. A bad system of taxation, encroaching upon the advances of the Cultivators. *Noli me tangere* is the motto of these advances.

2. Increase of taxes through expenses of collection.
3. Excess of luxurious expenditure on decoration.
4. Excess of expenses for litigation.
5. Lack of foreign trade in the produce of the land.
6. Lack of freedom in domestic trade in native commodities and in agriculture.
7. Personal vexations of the inhabitants of the rural districts.
8. Failure of the annual net product to return to the productive class.[13]

In Turgot, we find an additional, yet more subtle, argument for the redistribution of value in favour of the productive class of farmers. Let me recount the following passage already quoted above.

> The situation of the husbandman is materially different. The soil, independent of any other man, or of any agreement, pays him immediately the price of his toil. Nature does not bargain with him, or compel him to content himself with what is absolutely necessary. What she grants is neither limited to his wants, nor to a conditional valuation of the price of his day's work. It is a physical consequence of the fertility of the soil, and of justice, rather than of the difficulty of the means, which he has employed to render the soil fruitful.

Reading this passage in theological rather than simply economical terms, we can see how Nature grants the husbandman an amount of produce exceeding the price of his labour. The surplus value thus introduced into the economy is ultimately a gift of Nature rather than an acquisition through exchange. Even though the husbandman does offer his toil in exchange for the produce of the soil, this must be regarded as a token of gratitude rather than a service of equal value to the necessities of life provided by Nature. It is worth noticing that Nature is referred to by the feminine personal pronoun 'she'. What Mother Nature grants is not limited by the wants of the husbandman or even the price of his work. Her provision of the fruits of the soil seems to be almost an act of unconditional love. The theological undercurrent in physiocrat economics is that divinity manifests itself through the soil. Divinity is real. It is tempting to extend the analysis along Weberian lines to suggest that the primacy of the productive class of farmers is not only that they work the land to harvest the gifts of Mother Nature but also that this class seems to be the particular object of her divine love since this is the class to which she has decided to dedicate her gifts. When the physiocrats call for tax reform and free trade, this is to allow the value of these gifts to flow back to the class of people for whom they were originally intended.

The critique of the proprietary class's excessive appropriation of the value originally produced by the productive class of farmers rests on the proposition that the value of the work of the husbandman is mispriced in so far as he is not fully compensated for his provision of the produce flowing into the rest of the economy. Or perhaps the symbolic price of agricultural commodities does not fully represent their real value within the economy. This critique may be extended as we shift our perspective from the flow of value and money between the classes within the economy towards the interactions between the whole of the economy and nature as such. The economy as a whole is a net receiver of value from nature and we can hardly say that anything is offered in return. Value in the form of substance is rather a gift of nature. Physiocracy literally means the rule of nature. In Quesnay and Turgot, physiocracy is both a descriptive account of the economy and a normative ideal for its organization, an economy founded on the real value of nature.

A crucial concern in the writings of Quesnay as well as of Turgot is how to organize the economy in order to optimize the production of valuable output. In this sense, the physiocrats provide perhaps the very first theories of economic growth. Even though value ultimately comes from nature, the physiocrats are well aware that the fruits of the land do not emerge by the grace of Mother Nature alone. As we have already touched upon, the land must be cultivated in order to reap and increase the produce that it offers.

> Every species of labour, of cultivation, of industry, or of commerce, require advances. When people cultivate the ground, it is necessary to sow before they can reap; they must also support themselves until after the harvest. /.../
>
> The earth was ever the first and the only source of all riches: it is that which by cultivation produces all revenue; it is that which has afforded the first fund for advances, anterior to all cultivation. The first cultivator has taken the grain he has sown from such productions as the land had spontaneously produced; while waiting for the harvest, he has supported himself by hunting, by fishing, or upon wild fruits. His tools have been the branches of trees, procured in the forests, and cut with stones sharpened upon other stones; the animals wandering in the woods he has taken in the chace, caught them in his traps, or has subdued them unawares. At first he has made use of them for food, afterwards to help him in his labours. These first funds or capital have increased by degrees. Cattle were in early times the most sought after of all circulating property; and were also the easiest to accumulate; they perish, but they also breed, and this sort of riches is in some respects unperishable. This capital augments by generation alone, and affords an annual produce, either in milk, wool, leather, and other materials, which, with wood taken in the forest, have effected the first foundations for works of industry.[14]

We find here a very immediate reflection on the relation between investment and production, which is a key issue in subsequent theories about economic growth. The cultivator must sow before he can reap, and in order to sow he must conserve a portion of the previous harvest to be used as seeds. In more general economic terminology, the cultivator must allocate a portion of his current production required as capital investment to enable future production. If the cultivator or the other classes of society were to consume all of the produce generated in one year, there would be no investment to sustain or increase future production. Since agriculture is the only form of production capable of producing a surplus of value, which is required not only to support the other classes in society but also as capital in the development of the production capacities of the economy, the physiocrats recommend that as much value as possible flows back into the class of cultivators to facilitate agricultural capital formation.

As mentioned above, the physiocrats aim to ground their substance theory of value in an emphasis on the real dimension of the economy. This also applies to their account of capital formation. First of all, capital takes the form of actual physical matter, such as seeds, cattle, utensils for farming, buildings to hold the cattle and to store the produce, as well as food to support the cultivator himself and other members of his household until the next harvest. Second, the physiocrats include the idea of nature as the provider of the 'first fund for advance'. These first funds of capital constitute a gift of nature. Nature is the kick-starter of the economy providing an initial gift of the real upon which the symbolic order of economy may be founded. In the same way that a human being can never repay the initial gift of life that it was given at birth, so the economy can never repay the initial gift of the first advances of capital, which was provided at its inception.

Value as Labour

In his exploration of the principles of the market, where the price of commodities is settled, Adam Smith asks the following simple question: '[W]hat is the real measure of this exchangeable value; or, wherein consists the real price of all commodities.'[15] And he soon continues to provide an equally simple answer: 'Labour, therefore, is the real measure of the exchangeable value of all commodities. The real price of everything, what everything really costs to the man who wants to acquire it, is the toil and trouble of acquiring it.'[16] Essentially, this is the foundation of the labour theory of value found in classical economics. In the substance theory of the physiocrats, value is constituted by the organic physical matter that eventually becomes the object of human consumption. This is captured by the notion of *blé*. Given this material nature, value can be measured in terms of weight or volume: kilograms, tons, litres, barrels, bushels, etc. Obviously, labour itself cannot be measured in such terms. Even though we do sometimes resort to formulations, such as 'I have tons of work to do', strictly speaking this makes no sense. Instead, labour is measured in terms of time: hours, days, weeks, months or years. So according to the labour theory, the real value of a commodity is expressed through the amount of labour necessary for its production as measured in units of time. The difference between the substance theory of value and the labour theory of value is thus also a difference between matter and time. According to the labour theory, value is created as the labourer spends his time processing matter in a way that turns it into objects of human consumption. Labour is the inclusion of physical matter into the symbolic order of human desire. 'THE annual labour of every nation is the fund which originally supplies it with all the necessaries and conveniences of life which it annually consumes, and which consist always either in the immediate produce of that labour, or in what is purchased with that produce from other nations.'[17]

The relation between the conception of the economy found in the French physiocrats and the one found in English classical economists such as Adam Smith and David Ricardo is one of both continuity and discontinuity.[18] Both positions share the vision that free trade and less political intervention in markets serves to improve the overall productivity of the economy, which is potentially beneficial to all members of society. But the classical economists reject the idea of the agricultural cultivators as the only productive class in the economy, and Smith regards the physiocrats' designation of 'the class of artificers, manufacturers and merchants, as altogether barren and unproductive' as the 'capital error' of their system of thinking.[19] Instead, all forms of labour should be viewed as principally equal contributors to the wealth of the nation, regardless of whether they are employed in agriculture or industry.

The difference between the physiocrats and the classical economists marks a philosophical shift of emphasis from the domain of the real to the domain of the symbolic. This is a shift from a substance theory of value to a labour theory of value. The physiocrats' argument that the value created in agriculture has primacy over the value created by craftsmen and industry is ultimately based on an ontological postulate. Since nature is the origin of and precondition for all value creation, the class of cultivators is the most important one as they are the ones closest to nature. On this basis, the physiocrats dismiss their contemporary system of pricing, where cultivators receive only enough compensation for their produce to sustain their current level of production. In other words, the symbolic system of pricing is wrong. The physiocrats believe that if markets are deregulated and free trade is allowed, the system of pricing will be corrected and cultivators will receive a better price for their produce. We find here the contours of a fantasy of the rational market, which has the capacity of valuing commodities at the price corresponding to their true value. This phantasmatic image of the free and rational market opens up a new flank in the philosophical constitution of the physiocrats' system. What if free trade also benefits the allegedly sterile class of artisans and merchants? What if free trade also leads to a correction of the price of manufactured goods, allowing the class of artisans and merchants to produce and accumulate surplus value and perhaps even increase their volume of productive output? If this were the case, it would be difficult to reconcile an insistence on the primacy of agricultural production over manufacturing and industry while at the same time believing in the rationality of pricing in free markets.

In the context of eighteenth-century France, where the economy was excessively regulated and controlled in order to maintain the privileges of the landed aristocracy at the expense of a large population consisting primarily of farmers, this philosophical contradiction of physiocrat economics may very well be looked upon as a minor detail. The fantasy that truly free markets would adjust prices to the actual value of produce was probably easy to maintain in the historical context in which the physiocrats were writing as actual prices were on the whole not determined by free-market forces. But once the theory is generalized and applied in the understanding of economies with an increasing amount of industrial production, the significance of this detail is obviously perpetuated. This may serve as an explanation as to why the physiocrats' economic thinking was insufficient for the understanding of eighteenth- and nineteenth-century England, as the country was experiencing its transformation into the age of industrial capitalism. The intellectual shift performed by Adam Smith in order to understand this transformation consists in putting the human subject at the centre of the creation of value. As illustrated by the above quotation from the introduction to *The Wealth of Nations*, it is labour that 'originally supplies [the nation] with all the necessaries and conveniences of life'. Labour rather than nature is the true source of value in classical economics.

At this stage, it is important to note that the overall purpose of our eco-analysis and our specific reading of classical economics is not an exact representation of this school of economic thinking. There are major differences between different proponents of classical economics, and even an author such as Smith is not always wholly consistent throughout his writings.[20] Nevertheless, we shall be emphasizing those elements of Smith and Ricardo which are the clearest expressions of a pure labour theory of value.

A key difference between value creation by nature and value creation by human labour is that the latter is principally compensated through equivalent exchange while the latter is not. Through wages the labourer is compensated for his labour while there is, as we have already seen, no way to repay nature for the produce provided as input into the economy. In so far as the physiocrats still recognize the uncompensated contribution to the economy by nature, which is provided as a gift, they maintain that value is ultimately priceless. This understanding is in line with Žižek's definition of the real as that which 'resists symbolization'. The produce provided by nature is ultimately a form of value, which resists pricing. The gift from nature is priceless.

An often quoted slogan from Lacan reads: 'The letter kills.'[21] It refers to the way that our immediate access to the domain of the real is barred as soon as the process of symbolization sets in. As soon as the real has been integrated into a particular symbolic order of names, words and meanings, its immediate being is lost. It is no longer real but rather an integrated part of our perceived reality. In other words, the letter, or the symbol, kills the real. This is what happens when the real of the eco is integrated into the symbolic order of the economy. Economy functions as a form of eco-naming that has the effect of 'killing' the real of the eco. When classical economy attributes value creation to the performance of human labour, the origin of value in the real of the eco is veiled. The human subject of labour takes the whole credit for the gift of nature. Let's take a simple example.

A farmer spends ten hours of labour growing 100 kg of carrots that he brings to the market and sells for £100. As the carrots are sold, their value is priced. The eco is symbolized by the sign £. Within the framework of classical economy, this pricing has the effect that the entire value of the carrots are attributed to the labour of the farmer. In economic terms, it is as if the carrots have emerged *ex nihilo* as the fruits of the farmer's labour. Through his labour, the farmer has created new value at the amount of £100. What is effaced is of course the contribution of nature in the process of growing the carrots. Within physiocrat thinking, this contribution is still recognized as a gift, a unilateral contribution from nature to the economy. But within the labour theory of classical economics, the contribution from nature is concealed and in turn credited to the effort of the labouring subject. It is interesting to note how the concept of labour is also used in the context of giving birth to a child. In the physiocrats, we have seen how Mother Nature imbued with the fertility to produce the value, which is necessary for the reproduction of the economy. With the labour theory of value, it is as if the capacity to 'give birth' to new value is now displaced to the human subject of labour.

The mechanism by which the real of the eco is 'killed' as value creation is attributed to human labour also plays out in the classical understanding of capital and capital formation. This is most clearly expressed in Ricardo: '[T]he exchangeable value of [...] commodities produced would be in proportion to the labour bestowed on their production; not on their immediate production only, but on all those implements or machines required to give effect to the particular labour to which they were applied.'[22]

Few commodities are produced by labour alone as production involves the use of capital in some form or another. Still, the value of this capital is derived from the labour embodied in the production of this capital. If it takes two hours of labour to produce a hammer, the value of this hammer corresponds to the value of two hours of labour. The production of capital typically involves some form of capital itself. The production of a hammer would probably involve the use of already existing capital in the form of tools such as a plane, a saw or an anvil. The labour embodied in this capital also goes into the value of the hammer. This means that we can trace the value of commodities as well as capital back to some form of original labour. Ultimately, all value derives from labour. In this way, the labour theory of value effaces any kind of value derived from nature.

This principle also applies to the classical understanding of the value of land. Indeed, Ricardo speaks of the 'original and indestructible powers of the soil'.[23] The value of the land is priced in terms of the rent which the farmers pays to the landlord for permission to use the land. The landlord must leave enough produce for the farmer to maintain himself and his household and also to maintain the capital required to farm the land, but any produce over and above this may be appropriated by the landlord as payment for the use of his land. This idea of rent and re-investment was already a key point for the physiocrats but Ricardo takes a different turn as he develops a theory of the pricing of land:

> Nothing is more common than to hear of the advantages which the land possesses over every other source of useful produce, on account of the surplus which it yields in the form of rent. Yet when land is most abundant, when most productive and most fertile, it yields no rent; and it is only, when its powers decay, and less is yielded in return for labour, that a share of the original produce of the more fertile portions is set apart for rent.... If air, water, the elasticity of steam, and the pressure of the atmosphere, were of various qualities; if they could be appropriated, and each quality existed only in moderate abundance, they, as well as the land, would afford a rent, as the successive qualities were brought into use. With every worse quality employed, the value of the commodities in the manufacture of which they were used, would rise, because equal quantities of labour would be less productive. Man would do more by the sweat of his brow, and nature perform less; and the land would be no longer pre-eminent for its limited powers.[24]

The price of land in the form of rent emerges only when land becomes scarce. As long as there is a surplus of fertile land not yet appropriated, farmers have the opportunity to cultivate new land rather than pay rent to use land that has already been appropriated by a landlord. It is only when land becomes scarce or when land of only secondary quality is available for new appropriation that landlords are in a position to charge rent for the use of their land. The price of land that is charged as rent thus represents the value of the labour saved by farming an already appropriated piece of high-quality land compared to farming a piece of 'free' land of a lower quality, which requires more labour in order to provide the same yield. The price of produce is determined by the amount of labour required to produce it on those pieces of land that have been appropriated last. Landlords already sitting on the best and most fertile pieces of land thus profit as the economy expands to appropriate land of lower and lower quality. When the marginal production of produce requires more labour, the price of produce increases, allowing landlords to increase the rent on the high-quality land already appropriated.

> The reason then, why raw produce rises in comparative value, is because more labour is employed in the production of the last portion obtained, and not because a rent is paid to the landlord. The value of corn is regulated by the quantity of labour bestowed on its production on that quality of land, or with that portion of capital, which pays no rent. Corn is not high because a rent is paid, but a rent is paid because corn is high.[25]

As we have seen in the discussion of the physiocrats, the pricing of nature is obscured because the contribution of nature to the economy is basically free. Nothing is offered in exchange for the provisions of nature, and they remain thus ultimately priceless. However, in Ricardo's theory of rent we find a seeming solution to this problem. While Ricardo still recognizes the overall contribution of nature to the economy as a free gift, his theory enables the pricing of land to be subsumed under the labour theory of value. Rather than considering the overall contribution of nature to the economy, Ricardo's theory of rent provides a pricing of particular parcels of nature through comparison of their relative value. This allows him to use labour as the standard against which the relative values of different pieces of land are measured. High-quality land differs from low-quality land in that the former requires less labour than the latter to provide the same amount of produce. The value of land is an inverse function of the amount of labour required for its cultivation.

With the shift from a consideration of the absolute value of land towards a measurement of the relative value of land, the relation between the eco and the economy is subsumed by the measuring of value in terms of labour. The extra-economical contribution of nature in the form of unilateral gifts seems to vanish from sight as concern is shifted towards the way that the produce from high-quality land requiring relatively small amounts of labour may be exchanged for produce from low-quality land requiring relatively large amounts of labour. The difference between the two sides of the exchange is evened out through the payment of rent to the landlord. Even though the landlord does not produce anything, his contribution to the economy is still measured and even justified in terms of labour. As the landlord makes a piece of appropriated land available for productive farming, he is in effect 'saving' the economy from the expenditure of an amount of labour equal to the difference between the labour required to farm this land compared to the labour required to farm the next new piece of land that would otherwise have to be appropriated. In the labour theory of value, the landlord is not himself a labourer but still he is a 'labour saver', and his contribution to the economy can be measured by the same yardstick as that of the actual labourer. Rent is the compensation for this contribution.

If we extend the logic of Ricardo's reasoning, the contributions of nature can be exchanged for the efforts of productive labour. As the demands of the economy grow beyond the provisions immediately provided by nature, this is compensated through the labour of the productive subject: 'Man would do more by the sweat of his brow, and nature perform less.' Again, we see how the relation between the eco and the economy is subsumed by the measurement of value in terms of labour. In the substance theory of value, nature is still the ultimate source of economic growth as it provides surplus value in the form of gifts to the economy. The eco is more than the economy. In the labour theory of value, the contributions from nature become exchangeable for the efforts of human labour. Man himself is now the source and yardstick of value. The law of the eco-nomy wholly subsumes the real of the eco.

The difference between the substance value theory of the physiocrats and the labour value theory that can be distilled from classical economy can also be conceived in terms of theology. With the physiocrats, divinity enters into the theory in the form of giving Mother Nature. Mother Nature is not part of the domain of economy but provides the substance that is valued within the economy. In classical economy, we find divinity in the shape of Smith's famous notion of the invisible hand that was discussed earlier in our analysis of ecology. The divine intervention of the invisible hand is very different from the intervention of Mother Nature. While Mother Nature is an actual force of production, the invisible hand plays a much more subtle role by guiding the principles of organization and distribution within the economy. The invisible hand functions to guide the behaviour of the individual according to the laws of the market, thereby aligning the interests of the individual with the interests of society. In philosophical terms, divinity here enters through the domain of the symbolic as it takes the shape of the law guiding the forms of exchange taking place in the market.

Adam's Fall and the Making of 'Economic Man'

The issue of economic growth in classical economy is of course closely related to the notion of the division of labour. Here is how Smith begins his chapter 'Of the Division of Labour': 'THE greatest improvement in the productive powers of labour, and the greater part of the skill, dexterity, and judgment with which it is any where directed, or applied, seem to have been the effects of the division of labour.'[26]

Smith's account of the economy is put into motion by his observation of the phenomenon of the division of labour, whereby production is broken down into separate, specialized tasks that are carried out by individual workers as part of a complex whole. The division of labour enables an extreme increase in the level of productivity due to three different circumstances: increase of dexterity of the individual worker; time saving as the worker does not have to move between different tasks; and facilitation of the invention and improvements of machinery.[27] Growth is achieved through division of labour, which is a reorganization of production redeeming the productive powers of the labouring subject. We might even say that; in the labour theory of value, economic growth is achieved through the organization and shaping of subjectivity. A basic principle in Žižek's thinking as well as in psychoanalysis in general concerns the way that subjectivity and desire is constituted in the encounter with the laws inherent in the symbolic order. These are moral, legal, religious or other kinds of laws. According to Žižek, the concept of law is key to the emergence of subjectivity. If we take eco-nomy to mean not only 'eco-naming' but even 'the law of the eco', we may understand how the subject is included in the economy by being subjected to certain laws. The relation between law and subjectivity is unfolded in Žižek's reading of the fall of Adam. We shall make a small detour by reviewing this reading in order to subsequently apply it to the constitution of subjectivity and economy.

When Adam was expelled from the Garden of Eden, it was because he let his desires outweigh God's prohibition against eating the apple. An immediate reading of the story may look like this: first there was Adam's desire for the apple, and then there was God's prohibition against eating it, then came the transgression and finally the expulsion. Following Žižek's notion about the relationship between subject, law, desire and enjoyment (*jouissance*), these four moments should, however, be thought of as simultaneous. Desire is constituted only through prohibition. Prohibition in turn is conditioned by its transgression. And the desire for the apple is possible only when Adam is in a state already pointing beyond the Garden of Eden. Desire is thus only possible at the moment at which Adam is expelled. By presenting the story of Adam's fall successively, the impression is created that Adam loses something when he is expelled from the Garden of Eden. The truth is, however, that only at the moment that Adam is expelled is that which is lost produced. Only at the moment of expulsion is *jouissance* produced as that which is lost.

The story of Adam's Fall illustrates the paradoxical constitution of the subject. On the one hand, the subject is constituted as a desiring subject, a subject missing something, a subject in pursuit of an object that may fulfil its lack and provide *jouissance*. On the other hand, the subject is constituted by a lack in capacity for complete satisfaction, for absolute *jouissance*. The subject is in pursuit of something that is basically impossible. Being a subject is being part of a symbolic order of signs, language, prohibitions, law and so on. The subject's reflexive relation to himself and to his own desires is made possible only through the symbolic order. The symbolic order makes possible the subject's verbalization of his own lack: 'I want an ice cream'; 'I need a new car'; 'I miss having a girlfriend'. The symbolic order is not a 1 : 1 representation of subjective needs that were already there beforehand. A kind of simultaneity is at play whereby desire, the object of desire and the representation of both are constituted in one and the same momentum.

On the one hand, the symbolic order enables the subject's identification of his desires with different objects within the order. But, on the other hand, it is also the symbolic order that founds the subject's constitutive lack of Being which is the very precondition for desire at all. The symbolization of desire is at the same time the condition of impossibility for the complete satisfaction of desire. In order to describe the constitution of the subject, Žižek uses the rather dramatic concept of *symbolic castration*. Here is how he illustrates the concept with the fall of Adam:

> [W]hen Adam chooses to fall in order to retain *jouissance*, what he loses thereby is precisely *jouissance*... Adam loses X by directly choosing it, aiming to retain it.... That is to say: what, precisely, *is* symbolic castration? It is the prohibition of incest in the precise sense of the loss of something which the subject never possessed in the first place. Let us imagine a situation in which the subject aims at X (say, a series of pleasurable experiences); the operation of castration does not consist in depriving him of any of these experiences, but adds to the series a purely potential, nonexistent X, with respect to which the actually accessible experiences appear all of a sudden as lacking, not wholly satisfying. One can see here how the phallus functions as the very signifier of castration: the very signifier of the lack, the signifier which forbids the subject access to X, gives rise to its phantom....[28]

In order to explore the role of law and prohibition in the constitution of the economy, we shift the focus back to the other Adam. In Smith's account of the division of labour, we see the contours of the imperative of economic growth that mark capitalism in general and our own era of capitalism in particular. The economy appears to be guided by a law stating that the productive powers of labour must be increased. The division of labour is the method through which this law is brought into effect. Moving forward into the second chapter, where Smith examines the causes that bring about the division of labour, we find the following explanation:

> THIS division of labour, from which so many advantages are derived, is not originally the effect of any human wisdom, which foresees and intends that general opulence to which it gives occasion. It is the neces-

sary, though very slow and gradual, consequence of a certain propensity in human nature which has in view no such extensive utility; the propensity to truck, barter, and exchange one thing for another.[29]

The importance of this formulation lies in the way that Smith connects the foundation of the economy with the nature of Man. Economy emerges as the unfolding of the division of labour and the division of labour in itself is nothing but the effect of the 'propensity to truck, barter, and exchange one thing for another' inherent in human nature. In Smith's account, we find the same narrative structure as we saw in the immediate reading of Adam's fall. First, there is the human being imbued with the propensity to truck, barter, and exchange one thing for another. And then, as an effect of human nature, the economy emerges as the unfolding of the principle of the division of labour. This narrative structure provides a kind of anthropological founding of economy as the economy comes to stand for nothing but the social realization of 'propensities' inherent in the Nature of Man. The Nature of Man functions as a *point de capiton* where the symbolic order of economy is connected to the real of the eco.

With eco-analysis, this relation between subjectivity and economy should of course be unpacked and conceived differently. Rather than going along with the move from subjectivity to the economy, we should explain how they become the product of each other. In order to understand this interrelation between subjectivity and economy, we need to include another dimension of the division of labour. Chapter IV, where Smith introduces the notion of money, begins with the following passage:

> WHEN the division of labour has been once thoroughly established, it is but a very small part of a man's wants which the produce of his own labour can supply. He supplies the far greater part of them by exchanging that surplus part of the produce of his own labour, which is over and above his own consumption, for such parts of the produce of other men's labour as he has occasion for. Every man thus lives by exchanging, or becomes in some measure a merchant, and the society itself grows to be what is properly a commercial society.[30]

What is established in this passage is that division of labour is not merely the division of production into specialized tasks but also a division between labour as such on the one hand and consumption on the other. The subject, integrated into the symbolic order of economy characterized by division of labour, is dependent on the economy for the supply of produce to satisfy the far greater part of his wants. This means that the relation between production and consumption comes to be mediated by the symbolic order of the economy. From the perspective of eco-analysis, the implications of this division between labour and consumption goes further than what is immediately suggested by Smith. Not only is the subject dependent on the economy as a system of production and distribution for the provision of produce satisfy his wants, the economy itself comes to play an active role in the very constitution of his wants. The division of labour not only shapes the structure of the production apparatus; it also carries the seeds of modern consumer capitalism. Our desires are shaped by the selection of commodities offered in the market rather than the other way around. Or at least the two dimensions are mutually interdependent.

Perhaps the most emblematic consumer good today is the iPhone. Twenty years ago, mobile phones were a rare phenomenon used only by certain professionals such as emergency doctors or construction engineers. Most ordinary people did not have a mobile phone and they got along fine without it. Today, however, even small children are able to develop strongly felt needs to use and own mobile devices such as the iPhone, and they find it difficult to imagine how it was ever possible to live without the kind of instant electronic communication and entertainment provided by the iPhone and similar machines. This is a classic example of the way that supply is able to create its own demand. It is curious to note how the iPhone is not only produced by the company Apple but also branded with the iconic image of an apple from which one bite has been taken.

The obvious reading of Apple's apple image is of course to point to the biblical reference of Adam having violated God's prohibition and taken a bite of one of the fruits of the Tree of Knowledge. In this reading, the logo represents the human capacity for and aspirations of knowledge and technology. While Adam's fall may be seen as the result of him giving in to temptation, his eating of the apple also constitutes an act of rebellion against the authority of God. This layer of meaning may also be read into the logo on the iPhone as Apple likes to see itself as the rebellious company that is always looking to break new ground and never sticks to established rules. Without dismissing these interpretations, I would like to propose another and much more straightforward reading of the phantasmatic meaning inherent in the Apple logo. Eating an apple is a very simple way of satisfying the basic human feeling of hunger. While we all know very well that our desires for Prada shoes, new Playstation games, and L'Oreal perfume are nothing but the reflection of effective corporate branding, smart marketing and social pressure, the desire for an apple seems to be founded in an honest and uncorrupted need for nutrition.

Hence the image of the bitten apple invokes the fantasy of a fundamental, immediate and original need. This is perhaps how we should understand the logo: behind all the branding, marketing and technological evolutions, which makes it almost difficult to avoid buying into the empire of Apple gadgets, the iPhone ultimately corresponds to a very real and immediate need. In this way, the iPhone and the Apple logo emblematically incarnate a fantasy inherent in the economy. Beneath the layers of branding, marketing, manipulation, financialization, exploitation and so on, the economy ultimately provides us with the things that satisfy our most basic needs. The problem with this fantasy is not that it is false. The problem is that it functions to blur the distinction between those parts of the economy that produce actual apples for actually existing needs and those parts of the economy that produce Apple gadgets, while at the same time producing a demand for those very same gadgets which was not there beforehand. We shall return to this issue in part III.

4

The Market Theory of Value

The Two Laws of Exchange

> The concept of a general economic equilibrium based on balance of supply and demand has from the first played a central role in theoretical economics. In its simplest form the situation can be described roughly in the following terms: In a free market the price of each commodity depends on the extent to which it is demanded by consumers. If at a given set of prices the demand for a good exceeds the available supply then its price rises thus causing the demand to decrease, while if supply exceeds demand the price will drop and demand will thereupon increase. By this mechanism it is supposed that prices will eventually regulate themselves to values at which supply and demand exactly balance, these being the prices at economic equilibrium.[1]

This quotation comes from a 1955 article entitled 'The Law of Supply and Demand'. The article itself is hardly remarkable as it merely formulates a principle that may be found in any number of journals and textbooks on economy in the era of neo-classical economics. This is the principle of general equilibrium, stating that inherent in free markets is the capacity to generate prices reflecting the balance between supply and demand of a given commodity. A price, of course, is the rate at which a commodity is exchanged for money. When commodities are exchanged for money at the price settled in the free market, we can say that two entities of equivalent value are being exchanged. Hence we can see the concept of market equilibrium as an expression of a general law of equivalent exchange. It is worth recalling once again, how *nomos*, which is at the etymological root of the post-fix '-nomy' in economy, may be taken to mean rules or law. Economy thus means something along the lines of 'the law of the eco'. In brief, neo-classical economics is the institution of the law of equivalent exchange as the dominant law of the eco.

Of course, the idea that prices in the market are determined through the interaction between supply and demand is not an exclusive component of neo-classical economics. Albeit in a more crude form, this idea is also found in classical economics or even in the physiocrats. What is distinctively new about neo-classical economics in comparison with prior forms of economic thinking is the belief that the determination of prices at economic equilibrium provides a solution to the underlying problem of value. As we have seen, the physiocrats as well as the classical economists are concerned with the problem of the fundamental origin of value: Where does value come from? What is the real price of commodities? And how should commodities be priced? And in both positions, this question remains an enigma driving their intellectual inquiry. In eco-analytical terms, physiocrat economics and classical economics are both symptoms of the traumatic gap between value and price. In contrast, the major accomplishment of neo-classical economics is the formation of an imaginary fantasy that allows for the suppression of this traumatic gap. This is the fantasy of the free and efficient market: the market that is in concurrence with the law of equivalent exchange. In Žižek's terminology, such fantasy is inherently ideological. Let us re-invoke his definition of ideology, already presented on p. 26 in his analysis of ecology:

> Ideology is not a dreamlike illusion that we build to escape insupportable reality; in its basic dimension it is a fantasy-construction which serves as a support for our 'reality' itself: an 'illusion' which structures our effective, real social relations and thereby masks some insupportable, real, impossible kernel.... The function of ideology is not to offer us some point of escape from our reality but to offer us the social reality itself as an escape from some traumatic, real kernel.[2]

In the context of neo-classical economy, the notion of the free and efficient market is such a 'fantasy-construction which serves as a support for our economic reality itself'. It offers us the economy as an escape from the 'traumatic, real kernel' of the eco. The purpose of the following analysis is to explain and unfold what this actually means. We may arrive at an initial idea of this traumatic, real kernel of the eco by contrasting the law of equivalent exchange with another law governing the functioning of economy. This is the law of impossible exchange as conceived by Jean Baudrillard:

Everything starts from impossible exchange. The uncertainty of the world lies in the fact that it has no equivalent anywhere; it cannot be exchanged for anything. The uncertainty of thought lies in the fact that it cannot be exchanged either for truth or for reality. Is it thought which tips the world over into uncertainty, or the other way around? This in itself is part of the uncertainty.

There is no equivalent of the world. That might even be said to be its definition – or lack of it. No equivalent, no double, no representation, no mirror. Any mirror whatsoever would still be part of the world. There is not enough room both for the world and for its double. So there can be no verifying of the world. This is, indeed, why 'reality' is an imposture. Being without possible verification, the world is a fundamental illusion. Whatever can be verified locally, the uncertainty of the world, taken overall, is not open to debate. There is no integral calculus of the universe....

This is how it is with any system. The economic sphere, the sphere of all exchange, taken overall, cannot be exchanged for anything. There is no meta-economic equivalent of the economy anywhere, nothing to exchange it for as such, nothing with which to redeem it in another world. It is, in a sense, insolvent, and in any event insoluble to a global intelligence. And so it, too, is of the order of fundamental uncertainty.[3]

We can understand Baudrillard's notion of impossible exchange as a paradoxical twist to the concept of equivalent exchange. While, on the one hand, the economy is the sphere of equivalent exchange where commodities are exchanged for commodities, commodities for money or even money for money, the sphere of economy itself cannot be exchanged for something else. The economy as a whole is, in the poetic word of Baudrillard, insolvent. This is one way of conceiving the traumatic real kernel of the eco.

The concept of impossible exchange is hardly self-explanatory. As is usually the case with Baudrillard, his enigmatic propositions seem to be designed to put our thinking to work rather than to provide us with unambiguous definitions. In the following, we are thus going to work out what is actually meant by the economy being subject to the law of equivalent exchange while at the same time being subject to the law of impossible exchange. We can also think about the relation between the law of equivalent exchange and the law of impossible exchange in terms of Žižek's distinction between the real and the symbolic. As we have already touched upon, economy emerges when the eco (*oikos*) is subjected to a process of symbolic 'naming' (*nomos*). In brief, economy emerges through the process of 'eco-naming'. The kind of naming taking place in the constitution of the economy is quantitative rather than qualitative. The medium of eco-naming is money, and the 'names' of entities and events are expressed in terms of prices. In other words, economy emerges as the real of the eco is subjected to the symbolic process of pricing. The symbolization of entities and events in terms of prices makes it possible for them to enter into forms of exchange structured in accordance with the law of equivalent exchange. Prices make it possible to compare the value of different entities and events and exchange them for one another according to their value.

As long as the symbolic process of pricing, which is constitutive of economy, takes place in a piecemeal fashion, it functions relatively unproblematically. As long as prices are attached to individual entities and events that are gradually subsumed under the law of equivalent exchange as commodities and services, the economy and the market seem to work smoothly. But once the operation of pricing is applied to any kind of totality, the logic breaks down. The pricing of the value of one hour of labour makes sense, but the pricing of the totality of a human life does not. Even if we were able to arrive at a price, to whom would we be able to pay the corresponding amount of money? The pricing of a single commodity makes sense, but what is the price of commodification? The pricing of one acre of farmland makes sense, but the pricing of the entire planet Earth does not. The pricing of insurance against the default of a debtor makes sense, but what is the price of insurance against the total breakdown of the world economy, and who would pay the premium? The pricing of totalities come up against the law of impossible exchange.

The shift from classical to neo-classical economics is marked by the so-called marginal revolution. The prices emerging in the market, when commodities or other assets are exchanged for money, are marginal prices as they refer principally to the last individual trade which has been executed. A house is valued on the premise that it would be the next house to be sold on the market and this principle of pricing is then generalized to all other houses. This means that every individual house in the economy is valued at current market prices as if it were the next one to be sold on the market. This focus on marginal prices functions to veil the law of impossible exchange. On the margin, the law of equivalent exchange may apply, but when we think about the economy as a whole, we come up against the barrier of impossible exchange. Here is how Baudrillard continues from the quote above:

> Any system invents for itself a principle of equilibrium, exchange and value, causality and purpose, which plays on fixed oppositions: good and evil, true and false, sign and referent, subject and object. This is

the whole space of difference and regulation by difference which, as long as it functions, ensures the stability and dialectic movement of the whole. Up to this point, all is well. It is when this bipolar relationship breaks down, when the system short-circuits itself, that it generates its own critical mass, and veers off exponentially.[4]

The interplay between the law of equivalent exchange and the law of impossible exchange becomes particularly pertinent in relation to the issue of growth. As the economy grows in terms of GDP or similar measures of output, more and more domains of the eco are incorporated into the symbolic order of the economy and thus subjected to the law of equivalent exchange. From the marginalist perspective of the economy, 'all is well'. The pricing of yet another piece of the eco, which may then become the object of exchange for money, does not challenge the fundamental distinction between price and value. But, at certain points, the growth of the economy reaches a boundary where some form of totality is invoked. The classic example is of course the issue of climate change, where the continued emission of CO_2 from the burning of fossil fuel is projected to cause a significant disturbance to the balance of the atmosphere, causing global climate change of unpredictable magnitude. What seems to be at stake here is not just the survival of humankind but also the survival of the whole of the economy. Economy comes up against the barrier of impossible exchange as it finds itself unable to answer questions such as: What is the value of humankind expressed in price? What is the value of the Earth expressed in price? And what is the value of the economy itself expressed in price? It is at the point where such questions are posed that 'the system short-circuits itself, that it generates its own critical mass, and veers off exponentially'.

The Fantasy of Neo-Classical Economics

The fantasy of the free and efficient market, where the value of a commodity is determined in an equilibrium between supply and demand and expressed through the price at which the commodity is exchanged for money according to the law of equivalent exchange, allows for neo-classical economics to displace the problem of value. As we have seen, Smith poses the following question: '[W]herein consists the real price of all commodities?' The neo-classical answer to this question would be to simply point to the market and say: 'The real price of all commodities is the price quoted in the free and efficient market.' The discipline of economics gives up on the task of determining the real price of commodities and in turn institutes the notion of the market as the big Other that knows what the real price is. The fantasy of the market allows for the conflation of value and price. Value *is* price.

The fantasy of the market not only provides a reconciliation of the traumatic gap between the real of value and the symbol of price. It also allows for a seeming solution to the dispute between the substance theory of value and the labour theory of value. The dispute is solved through indistinction. The market does not care where value comes from. There is a market for carrots and there is a market for cars, and they work according to the same general principles. The market will determine a price for each of the commodities according to the relations between supply and demand. In neo-classical economics, the difference between the substance theory of value and the labour theory of value is a difference that does not make a difference.

I propose the conception of value found in neo-classical economics as a market theory of value. In combination with the substance theory and the labour theory, the market theory of value completes our analytical triad of value theories. Since the market theory of value relies on the phantasmatic image of the free and efficient market, the neo-classical form of economic thinking tends to shift the emphasis towards the domain of the imaginary. Neo-classical economics is not a theory about the economy as it is. It is a theory about how the economy would be if certain assumptions were true. It is not a theory about the way that actual markets work. It is a theory about how markets would work if they were to conform to the neo-classical definition of being free and efficient.

A key thinker in the school of neo-classical economics, Milton Friedman, elaborates on the relation between economic science and 'reality':

> The ultimate goal of a positive science is the development of a 'theory' or, 'hypothesis' that yields valid and meaningful (i.e., not truistic) predictions about phenomena not yet observed. /.../ Viewed as a body of substantive hypotheses, theory is to be judged by its predictive power for the class of phenomena which it is intended to 'explain'. /.../ Truly important and significant hypotheses will be found to have 'assumptions' that are wildly inaccurate descriptive representations of reality, and, in general, the more significant the theory, the more unrealistic the assumptions (in this sense). The reason is simple. A hypothesis is important if it 'explains' much by little, that is, if it abstracts the common and crucial elements from the mass of complex and detailed circumstances surrounding the phenomena to be explained

and permits valid predictions on the basis of them alone. To be important, therefore, a hypothesis must be descriptively false in its assumptions.[5]

According to this view, the truth of an economic theory lies not in the correspondence between the assumptions of the theory and any kind of economic reality. The truth of the theory is measured on its predictive powers with respect to future phenomena. The point here is not to dismiss the possibility that genuine insights about the functioning of the economy may be gained within such a paradigm of thinking. Still, the perspective of eco-analysis draws our attention to another effect of the relation between 'reality' and the phantasmatic images inherent in neo-classical economics. Friedman's conception of the relation between hypothesis/theory and assumptions corresponds nicely with Žižek's account of the relation between desire and fantasy. Žižek's concept of desire and fantasy could be described by the following Friedman paraphrase: 'Truly important and significant desires will be found to have "fantasies" that are wildly inaccurate descriptive representations of reality, and, in general, the more significant the desire, the more unrealistic the fantasy (in this sense).' As already discussed, fantasy has the function of projecting various imaginary qualities and potentials onto objects, which come to structure our desires. The following definition elaborates on the role of fantasy in relation to desire:

> Fantasy is usually conceived as a scenario that realizes the subject's desire. This elementary definition is quite adequate, on condition that we take it literally: what the fantasy stages is not a scene in which our desire is fulfilled, fully satisfied, but on the contrary, a scene that realizes, stages, the desire as such. The fundamental point of psychoanalysis is that desire is not something given in advance, but something that has to be constructed – and it is precisely the role of fantasy to give the coordinates of the subject's desire, to specify its object, to locate the position the subject assumes in it. It is only through fantasy that the subject is constituted as desiring: *through fantasy, we learn how to desire.*[6]

If we view neo-classical economics as simply a science engaged in posing hypothetical predictions and subsequently testing these against an empirical reality, it appears as a genuinely 'positive science' as opposed to a 'normative or regulative science', which is the contrast invoked by Friedman. However, this is to ignore the effects of the seemingly innocent assumptions on which neo-classical studies rest. These assumptions function as fantasy as they support a kind of desire to achieve the situation where they would in fact be true. Through the assumptions of neo-classical economics, we learn how to desire. This is where the normativity of neo-classical economics lies. The theory projects an image of an economy that would actually work according to the principles laid out in its models. While neo-classical economists do not claim that this is in any way a correct description of the actual state of affairs, the image still has the effect of making such a state of affairs desirable. In vulgar Marxist terminology, neo-classical economics is the ideological superstructure of contemporary capitalism. In the terminology of Žižek, the assumptions of neo-classical economics gives the coordinates of the eco's desire. A list of the most fundamental of these assumptions includes the following:

- Markets are free and efficient.
- Prices correspond to the real value of goods and services.
- Personal well-being corresponds to economic prosperity.
- Nations experience perpetual economic growth.
- Economic growth is unconstrained by limits imposed by the natural environment.

The success of neo-classical economics can be explained through these assumptions that function as fantasies. Even in our current times of economic and ecological crisis, neo-classical economics still maintains its position as the hegemonic paradigm of economic thinking in academic research as well as in political planning and decision making. What makes the neo-classical approach attractive is not accuracy in terms of describing, predicting or even just understanding the world in which we live. Rather, the strength of neo-classical economics is that it offers us economic reality as an escape from the world. The economic reality of neo-classical economics is in many ways a very pleasant reality in so far as it is a world where the above assumptions hold true: markets distribute wealth according to merit and achievement. People become happier as they grow richer. And there are no limits to the amount of wealth and happiness that may be achieved. This is of course a simplified account of neo-classical economics but the fundamental point is here that the main asset of neo-classical economics is its inherent fantasies about the economy as they deliver us from the unpleasant and traumatic nature of the real of the eco.

Neo-classical economics is attractive to academics as it provides an economic reality which can be measured and modelled in terms of mathematical algorithms. Why would you abandon this fantasy that economics is in the same premier league of pure hard science as physics and mathematics, rather than being comparable to soft pseudo-sciences such as history, psychology, sociology or even anthropology? And for economic planners and political decision makers, neo-classical economics is equally attractive as it allows for modelling that predicts the outcome of different decisions. Rather than having to take full responsibility for interventions in a world that is ultimately unpredictable, politicians are offered an economic reality where the only problems that appear are ones that may be addressed by tweaking the existing system and where technocratic modelling already points out the optimal solutions to these problems. Why would you abandon the fantasy that politics is merely a kind of economic engineering, where it can be predicted what works and what does not work, rather than being a radically undecidable matter of ethics or even revolution? In brief, neo-classical economics offers us economic reality in a form that readily lends itself to measurement, modelling, prediction and planning as an escape from a traumatic real eco that does none of the above. Therefore, as Mankiw formulates it:

> The purpose of economic theory is to take a complicated world, abstract from many details, and express the key economic relationships in a way that enhances understanding. From this standpoint, the neo-classical model is still the most useful theory of growth we have. It will continue to be the first growth model taught to students and the first growth model used by policy analysts.[7]

In the following analysis, we shall not only be looking into the neo-classical growth model but also some of the 'many details' from which the model abstracts.

Growth and the Production Function

> All theory depends on assumptions which are not quite true. That is what makes it theory. The art of successful theorizing is to make the inevitable simplifying assumptions in such a way that the final results are not very sensitive.[8]

This is how Robert M. Solow begins his seminal 1956 paper *A Contribution to the Theory of Economic Growth*. In the same year, Trevor Swan published the paper 'Economic Growth and Capital Accumulation', which independently arrived at some of the same conclusions as Solow.[9] These papers form the origins of the so-called Solow–Swan production function, which is at the heart of neo-classical thinking about growth. Both papers proceed in traditional neo-classical fashion by outlining a number of assumptions that allow for a particular modelling of an aspect of the economy. This is Solow:

> There is only one commodity, output as a whole, whose rate of production is designated $Y(t)$. Thus we can speak unambiguously of the community's real income. Part of each instant's output is consumed and the rest is saved and invested. The fraction of output saved is a constant s, so that the rate of saving is $sY(t)$. The community's stock of capital $K(t)$ takes the form of an accumulation of the composite commodity. Net investment is then just the rate of increase of this capital stock dK/dt or $\dot{K}$, so we have the basic identity at every instant of time:
>
> (1) $\quad \dot{K} = sY.$
>
> Output is produced with the help of two factors of production, capital and labor, whose rate of input is $L(t)$. Technological possibilities are represented by a production function:
>
> (2) $\quad Y = F(K, L).$[10]

As we have seen, the physiocrats as well as the classical economists were aiming to provide an answer to questions such as: Where does value come from? How is value created? The equivalent questions underlying the marginalist perspective of neo-classical economics would be the following: Where does *more* value come from? How is *more* value created? In the Solow–Swan production function, Y represents the total output of an economic community. Y is the total value created in the economy. Y is the outcome of the interaction between the two input variables of capital K and labour L. As long as the function is fixed, increase in the total output comes about only through the increase in one of the two input variables. An increase in the amount of labour employed in the production generates such an increased output. However, labourers are also consumers. This means that an increase in the employment of labour, which is merely the result of an increase in the population of the economic community, does not result in an increasing volume of output per person. Since the assumption of full employment is another one of the components of the model,[11] there is not much to be gained from an increase in the variable L. More labourers just means more mouths to feed anyway. What interests us is an increase in output per capita, rather than an increase in the total volume of output. This naturally directs all attention towards the input variable capital K as the key to increasing output per person. While labour is indeed recognized as a source of value in the neo-classical market theory of value, capital is the prime source of *more* value.

Perhaps the most important symbol in the chain of signifiers constituting the neo-classical production function is not Y, L or even K. It is the '=' that connects the two sides of the equation. The symbol stands for 'equality' or 'is equal to'. With Heidegger, we can also say that '=' simply stands for 'is'. And from Heidegger we also know that the very concept of 'to be' tends to be overlooked. This is what Heidegger refers to by his sweeping statement of the *Seinsvergessenheit* (forgetfulness of Being) of western metaphysics. In the following, I shall be pointing to different forms of forgetfulness pertaining to the neo-classical production function. I do not mean to say that they amount to *Seinsvergessenheit* in the strict Heideggerian sense of the word. But I do claim that they have profound implications for our way of thinking about growth at a time that is dominated by neo-classical economics. In terms of eco-analysis, we are going to see how the translation of the eco into the economy, the eco-naming of the eco, leaves crucial aspects of the eco behind. We are also going to see how these leftovers of the real come back to haunt the smooth functioning of the symbolic order of the economy. Again, neo-classical economics is the phantasmatic imaginary veiling the incongruence between the eco and the economy. In the same way that Žižek's psychoanalysis insists on breaking any notion of a self-sustained coherent subject down into the figure of the split subject \$, eco-analysis insists on breaking down the neo-classical image of the economy, $Y = F(K,L)$, into a notion of the split eco, $Y \neq F(K,L)$. Let us see how such playing around with symbols can make sense.

In equation (2), the '=' means that output equals input. You cannot get something out of nothing. As self-evident as such a proposition might seem, we shall see later how it functions to veil a fundamental component of economic growth. In equation (1), the '=' states that output equals savings plus consumption. This means that capital is increased as a portion of the productive output of the economy reserved for new investments, rather than being immediately consumed. The ability of the economic community to delay gratification increases their future capacity for consumption. An example typically given in economic textbooks is that of a farm, where a calf may be slaughtered and eaten or raised to become a grown cow producing milk. In the former case, productive output is consumed and, in the latter, productive output is invested. In the former case, the calf is immediately turned into a commodity. In the latter case, the cow becomes part of the economy's stock of capital. Given that the only source of additional capital is the productive output of the economy itself, the neo-classical production function tends to render the economy a closed system. This is also referred to as endogenous growth. Growth occurs when the amount of productive output that is recycled in the economy as investment exceeds the natural depreciation of existing capital.

The Solow–Swan model predicts that economic communities approach a so-called steady state, where the amount of output recycled in the economy as investment equals the amount required to compensate for the natural depreciation of the capital stock, as well as any expansion of the aggregate capital stock required to maintain a constant ratio of capital per worker in a situation where population growth increases the labour force. In the steady state, therefore, capital per worker is constant even if there is capital depreciation and population growth. The steady state represents a long-term equilibrium of the economy, where output per capita no longer grows given a constant savings ratio. The argument is that, as the capital stock of an economy grows, so also does the amount of capital lost each year due to natural depreciation. A larger share of the total investment is required in order to merely maintain the existing stock of capital and thus also maintain the existing level of output. Of course, the stock of capital may still be increased by increasing the savings ratio, that is, the ratio of investment relative to consumption. People may choose to forgo present consumption in order to invest in capital that increases future output. When the savings ratio is altered, the economy embarks on a new path approaching a new steady state, where a new equilibrium between investment, depreciation and population growth is approached. For every possible savings ratio of an economy, there is a corresponding steady state, where capital stock per worker is constant.

This notion of a steady state is further developed by Phelps in the formulation of his *Golden Rule of Capital Accumulation*. Phelps vividly presents his theory in the form of a 'Fable for Growthmen' in the imaginary kingdom of Solovia. Let us start by reviewing the assumptions behind his reasoning which are formulated as a 'committee report' prepared for the king of Solovia:

> The report expressed confidence that Solovia's supply of natural resources would remain adequate. It portrayed a competitive economy making full and efficient use of its only scarce factors, labor and capital, in the production of a single, all-satisfying commodity. Returns to scale were observed to be constant, and capital and labor were found to be so substitutable that fears of technological unemployment were dismissed.[12]

Again, who wouldn't want to live in such a place where there were unlimited natural resources, free markets and no unemployment? Phelps moves on from the Solow–Swan model to suggest that not only is it possible to derive a steady-state equilibrium for every possible savings ratio, it is also possible to derive a particular optimal savings ratio where consumption is maximized over time. The idea is to find a perfect balance between current consumption and future consumption. Rather than just urging people to save as much as possible, Phelps aims to find the optimal level of consumption and investment, where current generations consume at a level that allows for enough investment for future generations to consume at the same level but without current generations forgoing consumption at an amount that allows future generations to consume more. This optimum between current and future generations Phelps defines as the 'golden rule of capital accumulation': 'In a golden age governed by the golden rule, each generation invests on behalf of future generations that share of income which,... it would have had past generations invest on behalf of it. We have shown that, among golden-age paths of natural growth, that golden age is best which practices the golden rule.'[13]

In retrospect, there is something oddly confusing about the rhetoric used in the Solow–Swan notion of steady state, as well as Phelps's idea of the golden rule. Today, the concept of steady-state economy typically appears in the context of ecological economics, which implies a harsh critique of neo-classical economics.[14] On an immediate reading, one would think that the ideas of Solow, Swan and especially Phelps were fully aligned with the concerns of ecological economics that we need to organize our economy in a truly sustainable manner where current patterns of production and consumption do not compromise the life of future generations at the cost of the present. But of course, the devil is in the detail. In the case of the Solow–Swan model, as well as Phelps's elaboration of it, one of the details that has been abstracted away in order to make room for the elegant equations is the potential scarcity of land and natural resources. Solow himself notes how the composition of the equation 'amounts to assuming that there is no scarce nonaugmentable resource like land'.[15]

The Nomy Kills the Eco

An often quoted slogan from Lacan reads: 'The letter kills.'[16] It refers to the way that our immediate access to the domain of the real is barred as soon as the process of symbolization sets in. As soon as the real has been integrated into a particular symbolic order of names, words and meanings, its immediate being is lost. It is no longer real but rather an integrated part of our perceived reality. In other words, the letter, or the symbol, kills the real. This point is highly valid in the case of the economy. Once we appropriate conventional ways of economic thinking, everything seems to be conditioned by this particular symbolic order and we lose sight of that which is not captured by the economic symbolization. When we look at the way capital is recorded and incorporated in the neo-classical production function, we see how this procedure works to veil the real dimension of matter that eventually comes to constitute capital.

In the previous chapter, we saw how the labour theory of value views all capital as ultimately derived from labour. The buildings, machinery, tools and even the raw materials that are utilized in the production of commodities are all the products of human labour performed at some time in the past. The result of this thinking about capital is that the value of the contributions from nature to the production are effaced. The physiocrat notion of value as substance is wholly displaced by the idea of value as labour. A similar principle applies to the conception of capital found in neo-classical production. In the endogenous growth model of Solow and Swan, the economy grows as output is re-invested in the accumulation of capital stock. This accumulation then allows for an even larger production of output in the future, which in turn allows for more re-investment, and so on until the economy reaches a steady state where the maintenance of existing capital requires all of the new investments. This principle of growth and capital accumulation not only allows us to make projections into the future. It also accounts for the accumulation of capital that has been taking place in the past. Just as future capital stock is the result of current output being re-invested, so is all existing capital stock the result of past output that has been invested. This is a recent summary of the neo-classical understanding of capital: 'Capital, $K(t)$, represents the durable physical inputs, such as machines, buildings, pencils, and so on. These goods were produced sometime in the past by a production function of the form of [the Solow–Swan] equation.'[17] The inclusion of matter in the economy as capital is a process of symbolization. Machines, buildings and pencils are ultimately made out of iron, clay, wood and many other natural materials. As this matter is turned into objects designated as $K(t)$, the real of the eco is incorporated into the symbolic order of the economy. This is eco-naming. On immediate reading, the regress through which all capital is the product of previous production and investment seems to makes sense. Any machine, building or pencil was produced at some time in the past and their production was enabled only through the prior existence of other machines, buildings and pencils.

At the same time, the notion of capital as the product of prior production also has profound consequences for the distinction between what is inside and outside the economy. Let's take the example of a fishing vessel, which is part of a nation's capital stock. The fishing vessel consists of metal, wood, rubber and a number of other materials put together and then given a value expressed in money, either through its production cost or through its price at purchase. The purpose of our fishing vessel is to enable us to catch fish that eventually become part of the food production of the economy. But in order to catch fish, it is not enough just to have a fishing vessel. You also need a sea, where fish are continuously made available for catching through natural reproduction. In this sense, the sea, or perhaps more precisely the natural marine ecosystem of fish, is part of the production apparatus that creates an output in the form of food for human consumption. However, the sea itself was not 'produced sometime in the past by a production function', and therefore it is not recorded as part of a nation's capital stock.

We see here how there is an element of autism involved in the constitution of the economy. The conception of capital as an endogenous product of the economy itself systematically blocks out those forms of capital that are not produced by the economy itself. Figure 4.1 shows what the economy looks like according to such a conception of capital and production.

Variants of this model can be found in many a standard textbook in economics[18] and it reflects the same neo-classical understanding of economy that we have also found in the production function. The lower half of the diagram corresponds to the right-hand side of the production function, where we find the input factors of production. The upper half of the diagram corresponds to the left-hand side of the production function, which is where the goods and services produced are made available for purchase and consumption. As pointed out by Herman Daly and other scholars in ecological economics, such an image of the economy shows only the endogenous operations of the economy and fails to account for the exogenous interactions between the economy and the surrounding ecosystem.[19] The input factors accounted for in the image are only those that are already subject to scarcity and private ownership and thus incorporated into the pricing mechanisms of the economy. What is not accounted for is the energy and matter that is being extracted from the natural environment as part of the process of production.

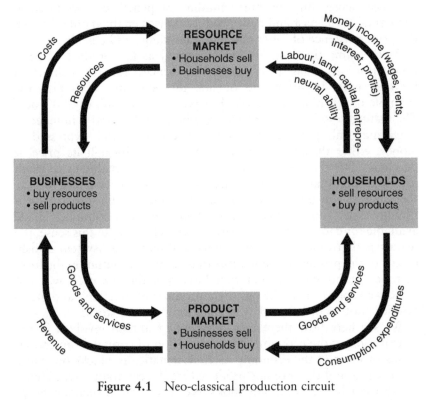

Figure 4.1 Neo-classical production circuit

The neo-classical theory of production only recognizes that which has been produced by the economy itself, as part of the economy. This means that a purpose-built fish farm, where fish are farmed in an enclosed environment created for that purpose through the use of machines and human labour, is recorded as part of a nation's capital stock according to its price of construction, whereas a natural lake capable of 'producing' the same amount of fish as the fish farm is not. The seemingly neutral recording of physical matter as capital thus functions to erase the natural conditions for production. In philosophical terms, the 'naming', or the nomy, functions to 'kill' the eco as we lose sight of the real in the process of symbolization. The production function only accounts for those forms of capital that are themselves produced thus leaving all 'unproduced' forms of 'capital' out of the equation. The sea or the lake, where fish come into being in a cycle of natural reproduction, are precise examples of an unproduced part of the real of eco. The list of other examples is as long as the number of natural ecosystems enabling the productive activities of human beings. It includes forests that provide timber, rivers that provide water for irrigation, topsoil that provides nutrition for plants, bees that provide pollination for plants and even the atmosphere that provides air for breathing.

This failure of the production function to account for substantial parts of the conditions for production extends to the very inputs that go into the goods that are finally made available for consumption. The production function not only fails to account for the sea that is the necessary condition for the existence of fish in the first place. It also fails to account for the fish itself. A very simplified application of the production function to the example of fishing looks like this:

$$Y(Fish) = F(K(fishing\ vessel), L(labour\ of\ fisherman))$$

In other words, the production of a fish that is ultimately eaten by a consumer is a function of the interaction of a fishing vessel and the labour of a fisherman. But where does the fish itself come from? Obviously, the business of fishing is not about making fish. Fishing is about moving fish from one place to another. You catch the fish as it is swimming in the sea and you bring it to a place where it can be bought for money by a consumer or by another business, such as a restaurant or a food producer that uses the fish as raw material in its own production. But even though any child would recognize that a fish originally comes from the sea, the production function manages to erase the origin of the fish. The fish appears on the left-hand side of the equation as the output of the production function, but it is nowhere to be found as an input in the equation. Since the fish is lacking on the input side, it looks as if the fisherman has not caught but simply created the fish through his labour. This illusion has profound implications for the way that neo-classical economics thinks, or perhaps does not think, about the relation between economic production and growth on the one hand and the natural environments in which this production and growth takes place. What neo-classical economics provides is a very convenient fantasy of the economy as a wholly self-sustaining system that is independent, or at least unconstrained by its natural environment. The limits to the growth of the economy are to be found within the economy rather than outside it. As long as the economy is capable of allocating a larger portion of the total output than what is required to compensate for the natural depreciation of the existing capital stock, it is capable of growing. Ultimately, this amounts to saying that as long as the economy as a whole is profitable, it is capable of growing. Did I forget to mention that the origins of neo-classical macroeconomics is neo-classical microeconomics? The neo-classical notion of growth is ultimately the application of a logic derived from the growth of individual companies on economic communities as a whole.

The mapping of the three different theories of value allows us to see how the three dimensions of Žižek's ontological triad, real–symbolic–imaginary, interacts with regard to the conception of capital. In the physiocrats' substance theory of value, the land is regarded as the ultimate source of value. The contribution from the land is conceived in terms of a unilateral gift and the value of this contribution cannot be priced in terms of money. In the physiocrats' conception, the economy is an open system that is dependent on an interaction with the non-economic real of the eco. The interaction between nature and the economy is not regulated by the law of equal exchange. Economy is in fact founded upon the law of impossible exchange. There is nothing which the economy has to offer in exchange for the fruits of the land provided by Mother Nature. The economy is primordially indebted to nature. With the classical labour theory of money, the value of nature's contribution to the products of the economy is conceived of in terms of the value of labour. While the farmer may not actually be able to produce the carrots with his own bare hands, the value of the soil may be measured in terms of the labour time saved by farming superior-quality land in comparison with inferior-quality land. Labour time provides the standard by which the value of land, produce and other natural resources is measured and symbolized.

If the eco-naming of the labour theory of value kills the real of the eco as uncovered by the physiocrats, neo-classical economics covers up the crime by hiding the body. Land and other natural resources do not even figure as original sources of value in the neo-classical production function. Only once they have been cultivated and turned into objects of capital do they become visible to the gaze of the market that translates value into price. The market theory of value is an imaginary fantasy of a self-sustained economy that owes nothing to anyone or anything. Everything that runs this economy and everything that is accumulated within this economy is a product of the efforts of this economy itself. This is because the market only sees things once they have already become part of the economy. The functioning of this imaginary of the economy may be further unfolded using our distinction between the two laws of exchange.

The Fish is Priceless

The reason why the fish in the above example is so obviously missing in the input side of the production function is because the equation has been filled out in qualitative rather than quantitative terms – in words rather than in numbers. This is a violation of the way that the equation is supposed to be used. Under normal circumstances, each of the variables in the equation is expressed as a figure denoting an amount of money or an amount of time. We would thus include the price of the fishing vessel and the amount of hours spent working by the fisherman. Putting these figures into the function, we would subsequently be able to equate them with the price at which the fish is ultimately sold. Since we would have numbers on both sides of the equation, the lack of the materially existing fish on the input side of the equation would not appear. The expression of the variables of the equation in terms of numbers turns the whole operation into an economic process that is subject to the law of equivalent exchange. Input is exchange for an output of equal value.

A price emerges when an object is exchanged for money in a market. The object may be actually exchanged or it may be imagined that the object could be exchanged and a price is generated from this imaginary exchange. In order for a market to function, there has to be a seller and a buyer. This means that someone has to own the object in order for it to be offered for sale on the market. If no one owns an object, it can be neither sold nor bought. The object cannot become subject to the law of equivalent exchange. This is why the sun does not have a price. No one owns the sun and therefore no one is able to sell it. The same principle applies to those parts of the earth that have not yet been appropriated through private ownership. And this is where we find part of the solution to the mystery of the missing fish. The problem with the fish is that no one owns it, as it swims into the fisherman's net. Therefore, the catching of the fish is not recorded as an exchange between a seller and a buyer and no price is generated. Second, there has to be some kind of scarcity for markets to work. If an object is abundant and readily available for everyone's consumption, the object will not appear as a commodity offered on a market in exchange for money. If you live on an island where there is an abundance of bananas and coconuts literally dropping down onto people's heads, there is not going to be a market for bananas and coconuts. Only when some of the land becomes enclosed as private property and the supply of bananas and coconuts is restricted to produce scarcity is it possible to establish a market where they may be sold and bought as commodities. We can also turn around the issue of ownership by saying that, prior to the enclosure as private property, the bananas and coconuts were owned by everyone.

Returning to the initial problem of the fish, the philosophical point is that it is ultimately priceless. The fish does not enter the economy at the moment it is caught by the fisherman. It only emerges in the economy when the vessel is back at the harbour and the fisherman sells it for money at the fish auction. Since the fish is not registered as input but merely as the product of the fisherman's labour, together with his utilization of the vessel, the value of the fish comes to be registered solely as output. This means that the symbolic order of the economy takes the whole credit for the very existence of the fish. The catching and selling of the fish counts as economic growth as if the fish had grown out of the economy in the same way that an apple tree grows out of the soil. What is not registered by the economy is that, prior to this transaction, the fish has disappeared from the body of things in the world that are not owned by anyone. The problem is that the fish does not have a price prior to being caught and the acquisition of the fish is not recorded as an exchange but rather as a unilateral appropriation.

Mainstream economic models of production and growth do not take into account the consumption of non-renewable natural resources such as oil, coal and gas, or the depletion of other natural resources such as fish stock, timber, fertile soil or fresh water through overexploitation and thus the undermining of the capacity of natural ecosystems to reproduce these resources. In this perspective, another more gloomy layer of meaning is added to our slogan that the 'nomy kills the eco'. As soon as natural resources are included in the symbolic order of economic exchange through the appropriation of ownership as well as valuation in price, their origin in the domain of the real, where neither property nor prices exist, is effaced. The act through which the nomy kills the eco has the character of a foundational violence in which market principles of equivalent exchange are violated. Depending on the way we think about nature, we can conceive of the inclusion of fish and other natural entities in the circulation of economy as a gift of nature or as a theft from nature. Gift giving and theft contradict the law of equivalent exchange governing exchange in the market, where commodities are exchanged for money assumed to constitute equal value. When oil is pumped out from under ground and into circulation in global capitalism, this is either a gift, given by *no one* or *anyone* to *someone*, or a form of theft, where the oil is taken by *someone* from *no one* or *everyone*. In either case, nothing is offered in exchange. Instead, we prefer to speak of the 'production' of oil as if the thing itself was made by economic activity.

Baudrillard writes about how the economy comes up against the barrier of impossible exchange because the whole of the economy cannot be exchanged for anything else. This principle holds not only for the economy but also for the eco itself. Ultimately, the eco cannot be exchanged. There is no one with whom the exchange can be made and there is nothing which could be given in return for the eco. We have already encountered Žižek's definition of the real as that which 'resists symbolization'. Since the eco is not originally owned by anyone, we can say that it resists pricing and it ultimately resists exchange. Obviously, this does not mean that the economy cannot put a price on natural resources or on the right to exploit natural ecosystems. Global commodities markets do precisely that. However, the point is that, in this symbolic process of pricing, there is a dimension of value that is not captured. There is something that is left over. This is the value that a natural resource or a natural ecosystem has for the sustainability of the eco itself. Since this value is left out of the pricing process, which only captures the value that the natural resource or the ecosystem has for the individual owner, it remains priceless. It is important to note here the double meaning of this word. On the one hand, to say that something is 'priceless' means that something does not have a price because it does not cost anything. In other words, it is free. This is true about the fish in our example. When the fisherman pulls it out of the ocean, he does not have to pay anyone. The fish is free. On the other hand, priceless also means that something is so valuable that the idea of putting a price on it must be written off as absurd. Some things have a value that cannot be expressed in money. This may not be true about the individual fish but it is true about a whole species of fish. Once a species has become extinct, it is gone for ever and part of a component in the ecosystem is lost. This loss is priceless. Again, the exchange of the eco comes up against the barrier of impossible exchange.

However, the ultimate impossibility of the appropriation of the eco by the symbolic order of economy not only constitutes the limit for the proliferation of economy, the ultimate limit of growth. It is at the same time the very condition of possibility for the economy. As noted above, a precondition for the establishment of economic exchange is scarcity. As long as objects and services are abundant and readily available for everyone, it makes no sense to enter into systems of equivalent exchange in order to attain them. Why buy coconuts at a shop if you can pick them up for free on the beach? In this context, the incorporation of the eco into the symbolic order of the economy and the possible destruction or deterioration of ecosystems by economic exploitation may become productive for the expansion and operation of the economy as it turns abundance into scarcity. An example is provided by the phenomenon where in some places around the world bees have become almost extinct as the result of the use of certain pesticides. Hitherto, the 'work' provided by bees in the pollination of crops was abundant and provided 'for free' as part of the functioning of natural ecosystems. As the bees die out, their services become scarce, but at the same time this creates a new market for pollination done by humans. Some companies are even doing research into the invention of robots that would be able to do the pollination previously provided by living bees. The destruction of ecosystems thus provides jobs for humans or even a new market for 'robobees' and ultimately economic growth.

The veiling of the origin of natural resources and the stock of natural capital in the reality of the eco takes place on the input side of the production function. But on the output side we can find a similar process working in the other direction. The production of many goods and services invariably generates waste products which are discarded in the natural environment. These may be by-products such as discharged water. They may also be emissions from the expenditure of energy used in the process of production, such as the emission of CO_2 from the burning of fossil fuels. Or they may simply be the commodities themselves, such as a computer that is turned into waste when the consumer has finished with it. But just like the input that is being extracted from the surrounding natural ecosystems, this flow of output back into natural ecosystems is also not recorded in the inherent pricing mechanisms of the economy. In similar fashion to the fish, waste is also priceless as it constitutes a real that resists pricing. Waste discarded into nature is a form of negative capital that is passed on to no one or anyone without any form of compensation. It is a kind of unwanted, negative gift. Again, we are dealing with a form of asymmetric exchange that eludes the pricing mechanisms of the market economy.

Of course, the preceding argument is very much in line with the critique of neo-classical economics raised by ecological economics.[20] As an alternative to the standard economics image of the economy, ecological economists provide another image that includes the economy as a subsystem of a larger ecosystem or biosphere. Figure 4.2 is one such image, taken from a course module in environmental economics.[21]

Comparing this model with the neo-classical textbook model presented in the previous section, we can recognize the flow of labour and capital from households to firms and the flow of goods and services going in the other direction. Both flows are mediated by payments of money going in both directions. However, the latter diagram also shows how energy and natural resources flow from the biosphere

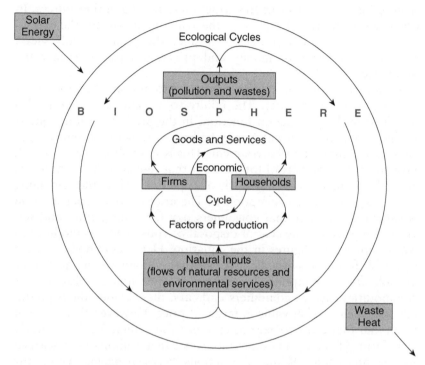

Figure 4.2 Ecological economics production circuit

and into the economic cycle, while waste and pollution is flowing from the economy and back into the biosphere.

The reason why the economy, when we look at it from the perspective of neo-classical economics, seems to be a perfectly closed circuit where output appears to be regenerated and reused as input, is because the symbolization process of the economy is so successful in confusing capital and commodities with their symbolic expressions in money. The real of the eco is translated into the symbolic order of economy through the process of pricing, thus effectively leaving out of the equation anything that is priceless. This means that when we look at the diagram in Figure 4.1, the economy seems to be nothing but a form of equivalent exchange between firms and households. When households, according to this neo-classical diagram, provide the 'markets for factors of production' with land and capital, they do not provide the market with the real natural resources required as input in the production of commodities. What households provide is the right to use land and investment capital in the form of money required in order for firms to extract these real natural resources. In other words, households provide the symbolic means for the extraction of real natural resources. But since the economy only views the world in terms of money and prices, it fails to register the difference.

A similar confusion applies to the consumption of commodities and the creation of waste. The failure to account for waste is also inscribed in the production function. In the production function, as we have seen, total output Y may be broken down into two components: consumption and investment. This is symbolized as: $Y = C + I$. Investment I is the part of total output that re-enters the economy as an addition to the existing stock of capital K, while the remaining part is expended as consumption C. The price at which consumption is valued is the price that consumers pay for commodities and services. No price for the cost of disposal of consumed commodities as waste into nature figures in the equation. Hence commodities seem to vanish from the symbolic order of the economy the moment they are sold for consumption. It is true that when households consume commodities such as computers and cars, these commodities gradually lose their market value in terms of price. However, this does not mean that they cease to exist as real material beings. They are merely transformed from commodities into waste that is ultimately discarded by the household. While transactions between agents within the economy are recorded and priced in different markets, there is no market for the pricing of transactions between the real of nature and the symbolic order of the economy. The absence of such markets is very obvious from the first of the two diagrams. Markets exist only for exchange between agents that are already inside the symbolic order of the economy. When we receive gifts from nature, when we steal from nature or when we violate nature through waste disposal, nothing is given back to nature and therefore no pricing takes place. The exchange between the eco and the economy remains priceless and it eventually runs up against the barrier of impossible exchange.

5

The Fantasy of Growth without Bounds

Knowledge or Technology

The initial formulations of the neo-classical production function in the 1950s by Solow and Swan includes only two input variables: capital and labour. As we have already discussed, an important implication of the models of Solow and Swan is their prediction that, over time, per capita growth of an economy ceases due to the phenomenon of diminishing returns to capital. In an economy with a low ratio of capital to labour, any additional investment in capital results in a relatively high increase in productivity and output. In a country without tractors, the first tractor introduced into the economy is going to make a huge difference in terms of productivity. But as the ratio of capital to labour increases, the marginal productivity gain for additional investments in capital decreases to a point where there is no benefit in new capital investments other than the ones compensating for depreciation of existing capital. There is only productive use for so many tractors per farm worker in an economy. At this point, the potential for further growth through capital investments is exhausted and the economy has reached the so-called steady state, where it grows only as a linear function of population growth.

The problem with this limitation on growth in the Solow and Swan models is the fact that it does not correspond very well with empirical observations of positive rates of per capita GDP growth in many countries persisting over significant periods of time.[1] Observing the long-term trends especially in western economies, economic growth does not seem to have limits. A solution to this problem was proposed by Paul M. Romer in the 1980s. In order to overcome the assumption of diminishing returns to capital leading to per capita growth rates approaching zero, Romer would introduce a third source of value that is not subject to such limitations. Here is how he presents his model:

> The model proposed here offers an alternative view of long-run prospects for growth. In a fully specified competitive equilibrium, per capita output can grow without bound, possibly at a rate that is monotonically increasing over time. The rate of investment and the rate of return on capital may increase rather than decrease with increases in the capital stock. The level of per capita output in different countries need not converge; growth may be persistently slower in less developed countries and may even fail to take place at all.... What is crucial for all of these results is a departure from the usual assumption of diminishing returns.
>
> [T]he model here can be viewed as an equilibrium model of endogenous technological change in which long-run growth is driven primarily by the accumulation of knowledge by forward-looking, profit-maximizing agents....In contrast to models in which capital exhibits diminishing marginal productivity, knowledge will grow without bound. Even if all other inputs are held constant, it will not be optimal to stop at some steady state where knowledge is constant and no new research is undertaken.[2]

The key point in this passage is the introduction of knowledge and technology as a third source of value creation that is not subject to the limitations of diminishing returns. The idea is that economic development is not merely driven forward by investment in more tractors but also by the invention of better tractors, as well as by the education of existing farm workers to become even better farm workers. While the marginal productivity gain of every additional tractor that is added to the economy is decreasing, the same does not apply to knowledge and technology. In the original growth model of Solow and Swan, additions to the stock of capital was a purely quantitative transformation of the economy. In contrast, improvements in knowledge and technology in the way conceptualized by Romer reconfigure the very relation between capital and labour. Such improvements create a qualitative transformation of the economy through transformations in the composition of capital and labour. While the addition of another old fashioned tractor to the economy is subject to decreasing marginal returns, the addition of a newly invented fully automated robotractor may transform the whole production mode of agriculture.

In the initial model of growth developed by Solow and Swan, knowledge and technology were either held constant or regarded as exogenous variables that were merely a cause but not an effect of economic development. But with Romer, knowledge and technology are included as an endogenous variable that is itself the outcome of surplus output being re-invested in the economy. Knowledge and technology become 'the basic form of capital'. The position of knowledge and technology in the production function is more recently explained in the seminal textbook on *Economic Growth* (2003) by Barro and Sala-i-Martin:

> In the real world, production takes place using many different inputs to production. We summarize all of them into just three: physical capital K (t), labour L (t), knowledge T (t). The production function takes the form:
>
> $Y(t) = F[K(t), L(t), T(t)]$
>
> Capital, K (t), represents the durable physical inputs, such as machines, buildings, pencils, and so on. These goods were produced sometime in the past by a production function of the form of [the above] equation. It is important to notice that these inputs cannot be used by multiple producers simultaneously. This last characteristic is known as *rivalry* – a good is *rival* if it cannot be used by several users at the same time.
>
> The second input to the production function is labor, L (t), and it represents the inputs associated with the human body. This input includes the number of workers and the amount of time they work, as well as their physical strength, skills, and health.
>
> The third input is the level of knowledge or technology, T... Workers and machines cannot produce anything without a *formula* or *blueprint* that shows them how to do it. This blueprint is what we call *knowledge or technology*.[3]

From an eco-analytical perspective, it is first and foremost curious to note how the definition of this third component, T, of the production function is itself hesitant and ambivalent. Already in Romer's exposition, the concepts of knowledge and technology are used interchangeably, but in this quotation the ambivalence is even more pronounced. The formulation: '[t]he third input is the level of knowledge or technology...' invariably begs the question 'So which one is it, knowledge or technology?' Strictly speaking, knowledge and technology are two different things. Why doesn't the quote simply say that T stands for knowledge *and* technology? This would immediately clear away the ambivalence. As it stands, the quote invokes a logic similar to the one found in the Marx Brothers' sketch frequently referred to by Žižek. In the sketch, Groucho is in a restaurant when the waiter approaches him with the standard question, 'Tea or coffee?' He replies, 'Yes, please!'[4] Groucho's reply is a refusal to choose between two options that are principally mutually exclusive. In similar fashion, the definition of T as 'knowledge or technology' functions to maintain an openness in the interpretation of the symbol.

The difference between knowledge and technology points back to the initial two input components in the production function. Capital, K, is exemplified as 'machines, buildings, pencils, and so on'. But what is a machine other than a piece of objectified technology? When the carpenter uses a car to transport material and equipment to and from the building site, he is using the technology embedded in the car as a machine. But the car already figures as capital in the production function. Even as simple a tool as a pencil incorporates an amount of technology. In other words, most of the things that we would consider to be capital are always already also pieces of technology. And vice versa: most of the things we would consider technology are always already capital. The same point applies with respect to labour, L, which is defined as including the labourers' 'physical strength, skills, and health'. But what is knowledge other than a skill? Especially when we think of knowledge in relation to production, it necessarily implies the dimension of 'knowing how to'. A skilled carpenter is a carpenter that knows how to install windows in a house. And, in general, a labourer that knows nothing is hardly a labourer at all. In the same way that a machine is a piece of objectified technology, the skills of a labourer constitute subjectified knowledge. Most of the capacities of a labourer that we would think of as skills are always already also knowledge. And vice versa: knowledge only becomes relevant for productions as it is translated into skills inherent in individual labourers.

Rather than seeing the ambiguity of the component T in the production function as an error or lack of clarity in the explication of the formulae, perhaps we should turn it around and understand the ambiguity as an inherent property of the notion of 'knowledge or technology'. If we regard labour and capital as the respective subjective and objective components of the production function, we can understand 'knowledge or technology' as a phantasmatic supplement to these two components. The input variable T seems to point towards some mysterious kind of *surplus* 'knowledge and technology' over and above the kinds of basic knowledge and technology that are already incorporated into labour and capital. To explore what this means, we return to the concept of *objet petit a*. In addition to the definitions already quoted, Žižek also offers the following account: '*objet petit a*, as the object of fantasy, is that "something in me more than myself" on account of which I perceive myself as "worthy of the Other's desire"'.[5] *T* has this character of *objet petit a* as it seems to stand for some unknown X-factor, for what is in the object of capital and the subject of labour more than capital and labour.

How can we tell the difference between a basic piece of capital and a special piece of capital imbued with the extraordinary properties of technology? From a purely 'objective' point of view, technology is invisible. Looking at a car, we can easily find the steering wheel, the engine and the rear-view mirror. But where do we find the technology? How can we tell if there is 'more technology' in a car than in a bicycle? The answer lies already in Žižek's definition of *objet petit a*. In the context of the production function, the market provides the ultimate proof of the productivity of capital. Technology is the *objet petit a* on account of which we perceive capital as worthy of the market's desire. There is more technology in a car than in a bicycle in so far as the car is more productive in terms of making goods and services that are ultimately demanded and purchased in the market. The *objet petit a* of technology is not immediately perceptible. We can only observe it indirectly as it is mediated by the desire of the market. Companies are willing to pay a higher price for a car than for a bicycle as an object of investment. We recognize the higher amount of 'technologiness' of the car as a retroactive effect of the higher price of the car. Technology is a phantasmatic property projected onto the object of capital through the gaze of the market.

The same applies to the subject of labour. How can we tell the difference between a simple labourer and a special labourer imbued with the extraordinary properties of knowledge? Again, we may easily locate the head, arms and feet of the labouring subject, but where do we find the knowledge of the subject? How can we tell if there is more knowledge in a graphics designer in Silicon Valley than in a factory worker in Zhengzhou? Again, the answer lies of course in the market. The superior amount of knowledge in the graphics designer reveals itself when his work adds more surplus value to the products he is producing. And this amount of surplus value is determined by the amount of money for which the product is exchanged in the market. This is similar to the process already described with respect to technology. With the inclusion of 'knowledge or technology' in the production function as an independent component, capital and labour are no longer just valuable in themselves. The variable *T* denotes the special capacity for bringing out that 'something in capital that is more than capital' and that 'something in labor that is more than labor'. The phantasmatic nature of 'knowledge or technology' means that their capacity is in principle restricted by nothing but the desire of the market. As long as the market is capable of mustering the desire for which 'knowledge or technology' are 'worthy', the economy may in principle 'grow without bound'. In what follows, we shall be looking into the relation between the 'knowledge or technology' and the desire of the market.

Gross Domestic Desire

We have already seen how, according to Žižek, 'the subject... and the object-cause of its desire... are strictly correlative.'[6] This means that our desire for a particular object is not merely a function of the inherent properties of the object but also a reflection of phantasmatic properties projected onto the object by the desire. When we transpose this psychoanalytic point into the field of eco-analysis, we see how it also applies in the relation between the two sides of the production function. The right-hand side is the side of production. This is where objects emerge as commodities and services. The left-hand side is the side of consumption. This is where these commodities and services become the object of desire. The two sides are 'strictly correlative'.

When capital and labour improve their productivity by being imbued with the magical elixir of technology and knowledge, this improvement is correlated with an equivalent increase in the desire of the market. In more straightforward terms, technology and knowledge are a source of value only if the surplus of commodities and services created through an increase in productivity are demanded by the market and exchanged for money by consumers. Essentially, production is nothing but the transformation of matter and action into objects and events that are desired by consumers. When someone first came up with the idea of integrating an MP3 music player into a mobile phone, this was regarded as productive progress since this new device became the object of consumers' desire, for which they were willing to pay more money than for a simple phone. Retroactively, the labour of the inventor would be ascribed a high level of knowledge. But if someone were to come up with the idea of integrating a hand blender and a phone, this object is not very likely to become the object of customers' desire and hence the labour of the inventor would not be ascribed any amount of knowledge. In fact, it would probably not even register as labour. Productive progress through knowledge and technology must be correlated with the desire of the market in order to register as economic growth.

We have already discussed how the development of Romer's model of growth is a shift from the purely quantitative understanding of growth in the Solow–Swan model to an understanding of growth as also a qualitative transformation of the economy. Romer's model is conceived to be a progress because it is capable of explaining why economies do not reach the state of steady-state growth predicted by Solow–Swan. While the two models differ with respect to their account of the production, they share an implicit assumption with respect to consumption. This is the assumption that the desire of the market is infinite. In principle, we could imagine that an economy would stop growing because people were happy and content with the amount of stuff that they already had. And in principle such a limit could be reached well before the productive capacity of the economy was fully exploited. This possibility does not, however, figure in either of the two models. It is assumed that, as long as production is capable of expanding the volume of output, the market has the capacity for absorbing more commodities and services. The assumption marks an important distinction between classical and neo-classical economics. In classical economics, we find much more elaborate reflections on the nature of desire and consumption. Instead of an infinite horizon of perpetual growth, John Stuart Mill speaks about the stationary state: 'It must always have been seen, more or less distinctly, by political economists, that the increase of wealth is not boundless: that at the end of what they term the progressive state lies the stationary state, that all progress in wealth is but a postponement of this, and that each step in advance is an approach to it.'[7] Even in Keynes, we find the idea that the current state of growth is merely a transitory stage on the path towards a future, where the 'economic problem' has been solved:

> Now it is true that the needs of human beings may seem to be insatiable. But they fall into two classes – those needs which are absolute in the sense that we feel them whatever the situation of our fellow human beings may be, and those which are relative in the sense that we feel them only if their satisfaction lifts us above, makes us feel superior to, our fellows. Needs of the second class, those which satisfy the desire for superiority, may indeed be insatiable; for the higher the general level, the higher still are they. But this is not so true of the absolute needs – a point may soon be reached, much sooner perhaps than we are all of us aware of, when these needs are satisfied in the sense that we prefer to devote our further energies to non-economic purposes. /.../ I draw the conclusion that, assuming no important wars and no important increase in population, the economic problem may be solved, or be at least within sight of solution, within a hundred years. This means that the economic problem is not – if we look into the future – the permanent problem of the human race.[8]

It is of course true that, at least with regard to capitalist societies, no economy seems to have reached a point of saturation where demand has stopped growing simply because people already have enough stuff for consumption. And even though we are today producing more stuff than ever before in the history of mankind, there are no signs that we should be approaching such a limit to the consumption capacity of the market. Along these lines, neo-classical economics seems to be correct in assuming that the capacity for consumption is limited only by the capacity for production. But rather than taking the insatiability of the market as an a priori condition of economics, we should include it as a problem to be explored within the analysis. At this point, there is a crucial difference between economics and eco-analysis. A standard textbook account defines economics as 'the study of how a person or society meets its unlimited needs and wants through the effective allocation of resources'.[9] Along these lines, we might define eco-analysis as 'the study of how a person or society is constituted to produce unlimited needs and wants through the effective phantasmatic projection of objects of desire'. If economic production theory is the study of the growth of gross domestic product, eco-analytical production theory also incorporates the study of gross domestic desire.

In the perspective of eco-analysis, the shift from the Solow–Swan to the Romer model can also be understood in terms of a shift in the constitution of the desire of the market. Knowledge and technology not only transform the production capacity of the economy but also work on the side of consumption by transforming the very desires of the consumer. In the textbook quoted above, we find the following definition: 'Workers and machines cannot produce anything without a *formula* or *blueprint* that shows them how to do it. This blueprint is what we call *knowledge or technology*.'[10] The attentive reader has of course already noted the striking resemblance to Žižek's definition of fantasy, which was also quoted earlier:

> The fundamental point of psychoanalysis is that desire is not something given in advance, but something that has to be constructed – and it is precisely the role of fantasy to give the coordinates of the subject's desire, to specify its object, to locate the position the subject assumes in it. It is only through fantasy that the subject is constituted as desiring: *through fantasy, we learn how to desire.*[11]

At first, this resemblance merely sums up what we have already discussed. Production is the transformation of matter and events into objects that are desired by consumers. Knowledge and technology constitute the phantasmatic dimension of labour and capital that is required in order for output to correspond to the desire of the consumer. The following paraphrase merges neo-classical growth theory with Žižek: 'It is only through knowledge or technology that labor and capital is constituted as productive: *through knowledge or technology, labor and capital learn how to produce.*' Fantasy operates in the order of the imaginary. Knowing how to produce is about being able to imagine the final product and to coordinate the process of production towards the realization of this product. 'It is the role of fantasy to give the coordinates of the production process, specify its object, and locate the position labour and capital assumes in it.' Knowledge not only enables labour to imagine the current object of production but even to imagine new objects of production around which the process of producing can be reorganized. Technology is the inscription of this imagination into the materiality of capital. Fantasy is an indispensable component of production. 'The fundamental point of Romer's model is that the form of production is not something given in advance, but something that has to be constructed.'

But, secondly, the resemblance between the two quotes also provides us with the key to understanding the relation between 'knowledge or technology' and consumption. Evolutions in knowledge and technology are not merely about optimizing the production of existing commodities or services or even about developing new commodities or services. Evolutions in knowledge and technology are also about expanding the scope of existing desires or even developing new desires in the consumer. 'It is only through knowledge and technology that the consumer is constituted as desiring: *through knowledge or technology, the consumer learns how to desire.*' This is of course a well-known fact in marketing theory and practice. It is not enough just to invent a smart new gadget. You also need to cultivate consumers into wanting and buying the thing. Nevertheless, this dimension of economic development is absent from neo-classical growth theory as it takes for granted the willingness and capacity of the market to keep absorbing more and new products and services. The insatiable consumer is inscribed into the model as an ontological fact.

In order to understand how knowledge and technology operate not only on the production side but also on the consumption side of the production function, we shall appropriate what Žižek refers to as Rumsfeldian epistemology. In 2002, Donald Rumsfeld gave a press release trying to justify the American invasion of Iraq, despite the absence of evidence that Saddam Hussein was indeed engaged in the development of weapons of mass destruction. Žižek subsequently picked up on Rumsfeld's distinction between different kinds of knowledge. In this passage, Žižek is quoting and commenting on Rumsfeld:

> 'There are known knowns. These are things we know that we know. There are known unknowns. That is to say, there are things that we know we don't know. But there are also unknown unknowns. There are things we don't know we don't know.' What he [Rumsfeld] forgot to add was the crucial fourth term: the 'unknown knowns', things we don't know that we know – which is precisely the Freudian unconscious. If Rumsfeld thought that the main dangers in the confrontation with Iraq were the 'unknown unknowns', the threats from Saddam the nature of which we did not even suspect, what we should reply is that the main dangers are, on the contrary, the 'unknown knowns', the disavowed beliefs and suppositions we are not even aware of adhering to ourselves.[12]

When Paul M. Romer speaks about 'knowledge as the basic form of capital', he is referring to knowledge, which enables new and more efficient forms of production. In terms of Rumsfeldian epistemology, this kind of new knowledge moves the boundary between 'known knowns' and 'known unknowns'. We currently know that we know how to manufacture a car that runs on petrol. We also know that we don't know how to manufacture a car that runs on water. In other words, the production of petrol-driven cars is a known known while the production of water-driven cars is a known unknown. But if someone were to come up with the knowledge to produce an energy-efficient car fuelled by water, the production of such a car would now be a known known rather than a known unknown. If we abstract from the fact that the invention of a water-driven car would be a massive catastrophe not only for the oil industry but probably also for much of the arms industry, as well as large sectors of the financial industry, the creation of such new knowledge would probably result in economic growth that would be recorded as an increase in GDP.

If, however, new knowledge is going to result in economic growth, it must also operate in the domain of the unknown. Henry Ford's following remark is famously quoted regarding his introduction of mass-produced cars to average consumers: 'If I'd asked people what they wanted, they would have said a *faster horse*.' What Ford is suggesting is not that people actually prefer faster horses to cars. In the usual interpretation, the point of the quote is that customers' demand is limited by their powers of imagination. Therefore, customers do not know what they *really* want. Only when they are presented with the actual new product are they able to recognize this as the true object of their desires. So people do want cars, they just do not know it yet. In terms of Rumsfeldian epistemology, the desire for cars, before they are marketed by Ford, is an unknown known. Following the straightforward interpretation of Ford, the art of marketing is about the conversion of unknown knowns into known knowns. The art of product development is about helping people to realize that they actually want a car rather than a faster horse.

But perhaps this conception of product development and consumption is too simple. While horses and cars are indeed two very different products, they both seem immediately to satisfy the same basic need for transportation. The development of the mass-produced car thus seems to be merely the invention of a new product that satisfies the existing demand for transportation in a more convenient and efficient way than the product already on the market, i.e., the horse. This was hardly the case even in Ford's time, and it is certainly not the case in the contemporary car industry. If the desire for cars was based on nothing but the simple need to move from A to B in the most efficient way, the industry would be looking a lot different and it probably wouldn't be the source of much economic growth. While cars do indeed provide consumers with the means to move from A to B, we also buy cars for a lot of other reasons of which we are only vaguely aware. Just like consumers in many other markets, the behaviour of the car customer is determined by known knowns as well as unknown knowns. This is of course a key psychoanalytic point. The desire of the subject is highly over-determined and anything but transparent to the conscious reflection of the subject itself. When shopping for a car, the consumer is thus subject to a whole set of 'disavowed beliefs and suppositions we are not even aware of adhering to ourselves'.

If the introduction of the mass-produced Ford T to substitute transportation by horse is conceived as a displacement of the desire for cars from the domain of unknown knowns to known knowns, much of contemporary economic growth is based on the displacement of desires from the domain of unknown unknowns to the domain of unknown knowns. The development of new products is not merely a matter of satisfying existing basic needs in more convenient and cost-efficient ways but also about the cultivation and creation of entirely new desires. If the Ford T was emblematic to economic growth in the age of industrial capitalism, perhaps the smartphone is equally symptomatic to contemporary consumer capitalism. The invention of the iPhone and similar devices does not merely provide smarter and more convenient means for communicating with other people. Aided by the myriad of so-called social media, they have brought about the neo-metaphysical world of the internet. Today, the desire for information and communication with other people seems to have collapsed into a drive to be constantly connected to this online world. Going back just one generation, this drive was located in the domain of the unknown unknown. In the 1980s, we did not know that we did not know that we wanted to be able to check our emails every half-hour and receive updates from our friends on Facebook every two minutes. Today, the drive to be online or perhaps rather the anxiety of being offline has moved into the domain of the unknown knowns. It is an impulse structuring our everyday behaviour on an almost habitual level, where we do not fully realize its inherent emotions and rationalities. We do not know that we know that we must be online. Nevertheless, this unknown known is the source of much consumer demand and thus economic growth.

In the previous discussion about economic castration, we have already explored the psychoanalytical intricacies in the way that desire is created as the subject is included in the symbolic order of the economy. The point here is that this relation between desire and economy is relevant in understanding what is missing from the neo-classical growth models. The shift from the Solow–Swan model to Romer's model that includes knowledge as the basic form of capital is perhaps symptomatic of a shift in the constitution of economic growth as such. It is true, as suggested by Romer, that knowledge and technology play a key role in growing contemporary economies in terms of output as recorded by GDP. Knowledge and technology bring about not only quantitative but also qualitative transformations of production. It is, however, equally true, as suppressed by Romer and the neo-classical conception of economics, that knowledge and technology also works to bring both quantitative as well as qualitative transformations on the side of the consumer. Product development and marketing go hand in hand. Increase in GDP is brought about as new products and new production methods are displaced from the domain of known unknowns to the domain of known knowns, but at the same time a concurrent increase in gross domestic desire (GDD) must be brought about by displacing the desire for new products from the domain of unknown unknowns to unknown knowns. When the transformation of the consumption side of the equation is excluded from neo-classical growth models, these models work to perform the ideological function of naturalization of desire. Rather than recognizing that economic growth not only increases the volume of commodities and services but also the desire for these commodities and services, neo-classical economics makes it seem as if economic growth is driven by an infinite consumer demand that is external to the economy. It fails to unveil how the knowledge of the consumer is also qualitatively transformed with the progress of economic growth.

This is another argument for the obvious fact that GDP is a poor measure of well-being. Regardless of the rate at which GDP increases, it never seems to catch up with the growth in GDD. Not only are the growth in the production of objects of desire and the desire for these objects, as Žižek would say, 'strictly correlative', but if we take standard measures of well-being as proximate expressions of the discrepancy between GDP and GDD, it even seems that any growth in GDP beyond a certain threshold results in an even bigger increase in GDD.[13]

Objective Demand and Consumptivity

On the right-hand side of the production equation, we find the distinction between capital and labour. As already noted, this distinction corresponds to the philosophical distinction between object and subject. Capital includes objects such as machinery, tools and raw materials, used in the production of output, and labour includes subjects facilitating this production through the interaction with the objects of capital. On the left-hand side of the equation, we do not, however, find an equivalent distinction. Total output Y may be broken down to two variables, consumption, C, and investment, I. The difference between the two variables is merely that C is the amount of output that is immediately consumed and I is the amount of output that is recycled into the production apparatus as new capital. While both objects and subjects are capable of creating productive output, only subjects are conceived of as capable of consuming output. The technical definition of consumption does not entirely correspond to the way that we usually understand the term. A commodity is recorded as consumption when it is exchanged for money without the purpose of facilitating further production. When I purchase a beer at the price of 15 kroner in the supermarket, the beer is now registered as having been consumed. C has increased by 15 kroner. However, the beer may be sitting in my cupboard for several months before I actually drink it. Or I may drop the beer on my way back from the supermarket without getting a taste of it. But none of this matters to the technical recording of the beer as consumption. Even though we normally think of the consumption of a beer as the actual drinking of it, the econometric conception of consumption is equivalent to the purchase of an object for money.

This is of course why only subjects have the capacity for consumption in the strict econometric sense. Only subjects have the capacity for ownership, which is the prerequisite for being able to buy things for money. A cat cannot buy anything because it cannot technically own anything. In fact, a cat cannot even own itself. Therefore a cat also cannot consume anything in the strict econometric sense. When the cat eats a bowl of cat food, this is registered as consumption by the human owner of the cat, who has purchased the cat food in the supermarket. The same applies even more obviously to objects such as bicycles, chairs or mountains. A bicycle cannot buy anything because it cannot own anything. We see also this subjective monopoly on consumption expressed by the fact that the GDP of a nation is typically measured on a per capita basis. Capita is derived from the Latin *caput*, which means head. Since we are obviously not counting the heads of cats, pigs and other animals, economic progress is measured by growth in the average amount of goods and services that is available for purchase by every human being in the particular economic community. In brief, subjective consumption is the ultimate measure of economic growth.

Although these points about consumption, cats and bicycles may seem like sophistry, they do in fact have implications for our understanding of growth. In the previous section, we have seen how the capacity for consumption in the economy is expanded through the quantitative as well as qualitative expansion of the subject's desire. The expansion takes place as new products and services are developed by new knowledge and technology. But even this conception of desire seems to be insufficient for understanding the way that consumption operates today. We cannot take for granted that only subjects consume. Objects also have the capacity for consumption. The move from subjective to objective consumption is comprehensible through Žižek's distinction between interactivity and inter-passivity:

> [T]hink about the canned laughter on a TV-show, when the reaction of laughter to a comic scene is included in the soundtrack itself. Even if I do not laugh, but simply stare at the screen, tired after a hard day's work, I nonetheless feel relieved after the show, as if the soundtrack has done the laughing for me.
>
> To properly grasp this strange process, one should supplement the fashionable notion of interactivity, with its uncanny double, *interpassivity*. It is commonplace to emphasize how, with new electronic media, the passive consumption of a text or a work of art is over: I no longer merely stare at the screen, I increasingly interact with it, entering into a dialogic relationship with it (from choosing the programs, through participating in debates in a Virtual Community, to directly determin-

ing the outcome of the plot in so-called 'interactive narratives'). Those who praise the democratic potential of new media generally focus on precisely these features: how cyberspace opens up the possibility for the large majority of people to break out of the role of the passive observer following the spectacle staged by others, and to participate actively not only in the spectacle, but more and more in establishing the rules of the spectacle.[14]

Along the lines of the previous analysis, interactivity is an example of the way that new knowledge and technology are used to expand the subject's desires. As we have already touched upon, smartphone technology combined with social media function to create and propel the subject's desire for online interactivity. But here is how Žižek continues:

> The other side of this interactivity is interpassivity. The obverse of interacting with the object (instead of just passively following the show) is the situation in which the object itself takes from me, deprives me of, my own passivity, so that it is the object itself which enjoys the show instead of me, relieving me of the duty to enjoy myself. Almost every VCR aficionado who compulsively records movies (myself among them) is well aware that the immediate effect of owning a VCR is that one effectively watches *fewer* films than in the good old days of a simple TV set. One never has time for TV, so, instead of losing a precious evening, one simply tapes the film and stores it for a future viewing (for which, of course, there is almost never time). Although I do not actually watch the films, the very awareness that the films I love are stored in my video library gives me a profound satisfaction and, occasionally, enables me to simply relax and indulge in the exquisite art of *far niente* – as if the VCR is in a way *watching them for me, in my place*. VCR stands here for the big Other, the medium of symbolic registration.[15]

At first, we may note that, even though the quote is less than ten years old, the example of the VCR seems hopelessly dated. With the explosion of new technologies allowing for video-on-demand, streaming or just plain download of films from illegal websites, few people ever use VCRs anymore. In itself, this is a fine illustration of the speed at which patterns of consumption are currently changing. But the real point is of course how Žižek invokes the concept of interactivity to account for the way that objects do not merely function as media for subjective consumption and enjoyment. Objects such as a VCR are even capable of enjoying for the subject. While the subject has only a limited amount of time available for watching television shows, the VCR greatly expands this capacity by being able in principle to record for 24 hours a day.

The example provided by Žižek immediately relates to the consumption of cultural products such as films and television shows but the idea of objects consuming for the subject may be generalized to also encompass the consumption of physical commodities. Arguably, one of the most important commodities in our contemporary economy is oil. The global consumption of crude oil is estimated at around 90 million barrels per day. Who or what actually consumes these vast quantities? Obviously, crude oil is not fit for direct human consumption in the non-econometric sense of the word. Instead, crude oil is refined and used to fuel cars, aeroplanes, ships, power plants and a whole range of other objects. While it is of course true that the ultimate purpose of these oil-fuelled objects is to provide goods and services for subjective consumption and enjoyment, it is also true that the development and introduction of these objects into the economy greatly expands the aggregate capacity for consumption. On average, every UK citizen consumes one-fortieth of a barrel of oil per day. This would not be possible without the aid of cars, aeroplanes and all the other objects that do this consumption for the subject.

The point here is not some vulgar pseudo-Buddhist idea that we should abandon all forms of technology and return to an original form of hunter-gatherer lifestyle. It is merely to suggest that economic progress through knowledge and technology does not necessarily translate directly into more commodities and services available for subjective consumption or even into more free time and spare money because the same amount of stuff can now be produced cheaper and with less labour. The development of new products also functions to produce a whole new set of objective desires demanding to be satisfied. The development of the car not only produced a desire for petrol but also for roads, vehicle repair, traffic police, ambulance services and so on. Another example is the way that the development of the Playstation game console produced a desire for new games. And the development of some of these games produced a desire for additional hardware, such as Playstation steering wheels for racing games, headsets, guns for shooting games and so on. The development of such new products not only operates on the supply side by increasing the amount of objects available for subjective consumption. The products themselves come to figure on the demand side as they also act as a kind of objective consumer.

Since subjective and objective consumption are intimately intertwined, it is of course impossible to measure their relation. When a family uses a car to go on holiday, the consumption of oil is perceived as part of the provision of the holiday as a service. And when parents buy Playstation accessories for their children, these are perceived as additional commodities enhancing the experience and enjoyment of the child. Still, we need to ask ourselves the impossible question: how much of our productive progress through knowledge and technology is absorbed by maintaining and servicing the surplus demands coming from this very same progress? With eco-analysis, we can think of the difference between objective and subjective consumption as the difference between the real and the symbolic. A key difference between objects and subjects is that only subjects have the capacity for enjoyment. This means that objective consumption is consumption without enjoyment. Think about a car running idle. Unless the driver gets some form of satisfaction from the sound, smell and feel of the engine running without the car moving, this is a purely objective form of consumption. Petrol is being consumed without corresponding to any form of subjective consumption. Now, idling is a limited case because it clearly does not correspond to any form of subjective consumption. The case of the family driving on holiday is different. In this case, there is still an objective consumption of oil but now there is some kind of correspondence with the enjoyment achieved through the subjective consumption by the family members. The objective consumption of oil by the car is integrated into the symbolic order of subjective consumption as it comes to figure as a means of satisfying the desires of the family members.

On the right-hand side of the production function, the efficiency of the economy is expressed in terms of productivity, which is a measure of the amount of capital and labour required in order to produce a certain amount of output. In an efficient economy, capital and labour yield a larger amount of output. However, we do not find an equivalent measure on the left hand side of the production function. What if economies differ not only with respect to their efficiency in converting capital and labour into commodities and services but also with respect to their efficiency in converting commodities and services into enjoyment? In order to account for such differences, we would need a measure of the *consumptivity* of the economy. The mathematical expression of such a measure would look like this:

$$CP = \frac{SC}{OC}$$

where:

CP = Consumptivity
SC = Subjective consumption
OC = Objective consumption

The point here is of course that when the ratio of subjective consumption relative to objective consumption decreases, the amount of enjoyment yielded by a given amount of output also decreases. From a standard econometric perspective, a major problem with this formula is, however, that OC is impossible to measure. If we follow the principles of neo-classical economics, we are forced to assume that all objective consumption is always already also subjective consumption. Let's imagine that our family on holiday misses an exit on the motorway and is forced to make a 20-kilometre detour, using an additional litre of petrol. By definition, the neo-classical model assumes that the family members have now increased their consumption by 2 euro or whatever the price is of the fuel. In other words, the value of the holiday has increased by 2 euro. This is because consumptivity is by definition fixed at a value of 1. Objective consumption is always already registered as subjective consumption. In the terminology of Žižek, the difference between objective and subjective consumption 'resists symbolization', and the neo-classical model thus performs the ideological function of veiling this difference through the fantasy that all forms of consumption are ultimately redeemed by the subject.

The issue of subjective and objective consumption is vividly illustrated by the so-called food versus fuel controversy over biofuels. The increasing production of fuel from maize, sugar cane, soy beans or other forms of plant material over the past decade is beginning to show significant consequences for the production as well as the pricing of food for human consumption. More and more farmland is being converted to the production of crops for biofuels, and the use of crops for fuel pushes up demand, which in turn results in price increases, as well as higher price volatility in agricultural commodities markets.[16] One reason why this issue has raised such controversy is perhaps because it spells out, in a very direct way, how cars compete with humans over access to the same resources. In eco-analytical terms, food versus fuel is a controversy between subjective and objective consumption. The issue also reveals how we have come to populate the world with machines that are literally taking the food out of the mouths of people.

The argument of the current, as well as the previous, section may be summarized in the formulation of the following hypothesis: over the course of the past four decades, increases in economic growth as measured in GDP per person, as well as in economic productivity as measured in GDP per unit of labour and capital in western economies, have been accompanied by a concurrent *falling rate of enjoyment*. The hypothesis is an elaboration of the following remark by Baudrillard:

Enjoyment is radical, value is sublime; so this radical symbolic insistence is sublimated in value. The commodity is the incarnation of the sublime in the economic order. The radical demand of the subject is sublimated there in the ever renewed positivity of his demand for objects. But behind this sublime realization of value, there lies something else. Something other speaks, something irreducible that can take the form of violent destruction, but most frequently assumes the cloaked form of deficit, of the exhaustion and refusal of cathexis, of resistance to satisfaction and refusal of fulfilment. Viewing the contemporary economic situation as a whole, all this begins to look like a tendency we might want to call the falling rate of enjoyment.[17]

Baudrillard himself does not further explicate the concept of the falling rate of enjoyment but our previous analysis provides us with the means to continue his line of thought. On the production side, the development of new knowledge and technology has functioned to increase the volume of output of goods and services available for purchase in the market. Obviously, the increase in production is mirrored by an increase in consumption. But, at the same time, the increase in productivity is also mirrored by a decrease in consumptivity. The development of new knowledge and technology has consequences for the constitution of the consumer subject so that new desires are created and GDD is increased. Furthermore, the development of new products through knowledge and technology also creates more and new objective consumers. The composition of consumption is changing so that the ratio of subjective consumption relative to objective consumption decreases. Since only subjective consumption results in enjoyment, the amount of enjoyment per unit of output decreases correspondingly. This is what we may refer to as the falling rate of enjoyment.

This tendency to a falling rate of enjoyment constitutes an inconvenient underside to the prospect of economic growth without the constraints of diminishing marginal returns to capital as put forward in Romer's fantasy of 'growth without bounds'. The introduction of knowledge and technology into the production function may overcome the limitations of diminishing marginal productivity, but the problem is merely displaced to the left-hand side of the equation, where knowledge and technology cause a diminishing marginal consumptivity. The amount of surplus enjoyment yielded by an increase in one unit of output is becoming gradually smaller and smaller. Fortunately, enjoyment is such an elusive concept that the hypothesis of the falling rate of enjoyment is beyond any kind of econometric validation or falsification, which makes the problem very easy to ignore.

GDP as the Gold Standard of Living

Let us just briefly recount the purpose of the current eco-analysis of neo-classical economics and its inherent conceptions of economic growth. Even if few people actually master the technicalities of the highly sophisticated economic models of neo-classical economics, the school of thought seems to be an eminent expression of the way that western societies think about growth, production, investment and consumption. Neo-classical economics is a very pure expression of the ideology of contemporary capitalism. This ideology not only structures the relations between subjects and objects in the world, such as the ones between economic agents and their natural environment, but also the way that western societies understand themselves and their place in history. Also, in this respect, the notion of knowledge and technology has a crucial function. In this chapter, we shall be further unfolding the eco-analysis of knowledge and technology while at the same time summing up some of the points of the previous analyses.

The following quotation comes from a standard textbook in macroeconomics, and it constitutes the introduction to the chapter on economic growth.

> If you have ever spoken with your grandparents about what their lives were like when they were young, most likely you learned an important lesson about economic growth: material standards of living have improved substantially over time for most families in most countries. This advance comes from rising incomes, which have allowed people to consume greater quantities of goods and services.
>
> To measure economic growth, economists use data on gross domestic product (GDP), which measures the total income of everyone in the economy. The real GDP of the United Kingdom today is more than three times its level 50 years ago, while real GDP per person has grown by only a slightly lower multiple (a factor of around two and three-quarters) over the same period. In any given year, we also observe large differences in the standard of living among countries. Table 7-1 shows income per person in 2005 for 26 countries (these figures are calculated by the World Bank, and, as is customary in international income comparisons of this kind, they are given in US dollars). The United States tops the list, with an income of $41,950 per person. Burundi has an income per person of only $640 – barely 1.5 per cent of the figure for the United States.[18]

The quote captures very well our immediate ideas about economic growth and the relations between growth and welfare. GDP represents the productive capacity of society and hence also people's capacity for the consumption of goods and services. Economic growth is aligned with progress and prosperity as the image of the poor old days is invoked: 'If you have ever spoken with your grandparents about what their lives were like when they were young...' Economic growth is what got our societies to the stage of civilization where we are at today. We no longer have to worry about going hungry because the harvest has failed. In fact, most of us do not even have to worry about the harvest at all since agriculture has become industrialized to the point where only a fraction of the population is involved in the production of primary food supplies. Thanks to economic growth, work is now something that takes place in front of a computer rather than behind a plough. At least, this is likely to be the case for the young people reading Mankiw and Taylor's economics textbook.

As we have already seen, knowledge and technology play a key role in the explanation of economic growth and the historical progress of western capitalist societies. We may recall this passage from the previously quoted definition of the variable T: 'Technology can improve over time – for example, the same amount of capital and labor yields a larger quantity of output in 2000 than in 1900 because the technology employed in 2000 is superior.' The idea of progress is intrinsic to the self-conception of modern capitalist states, so the idea that we are becoming constantly richer and richer because we are becoming constantly smarter and smarter has a very immediate appeal. As we have already formulated it in Žižek's terminology, T stands for *objet petit a*, with its capacity to bring out that in the object of capital and in the subject of labour which is more than capital and labour. Knowledge and technology not only pave the way for the fantasy of unbounded perpetual economic growth, they also provide a normative justification for this growth, whereby rich economies deserve their wealth since it has been achieved through their own intellectual superiority. The famous L'Oreal slogan could be used to capture such justification: 'You are experiencing GDP growth *because you're worth it!*'

When we explain sustained economic growth in an economy over a period of time through improvements in 'knowledge or technology', our explanation has the effect of simultaneously covering two very different phenomena. On the one hand, 'knowledge or technology' may function to optimize the production of existing products and services in a way that enables us to produce more of the same thing with less effort or less resources. This is the case, for instance, when engineers improve the technology in coal-power plants by which electricity or heat are produced from the burning of coal. We may refer to this as *vertical growth*. On the other hand, 'knowledge or technology' may also function to expand the sphere of the market economy, where goods and services are exchanged for money. This happens when the market economy takes over forms of production and exchange that were previously serviced by other forms of economy, such as a gift economy, a barter economy or an economy of sharing. When, for instance, the development of welfare technologies such as kindergartens and the education of pedagogical professionals transform child care from being a service provided as part of the gift economy of the family into a service produced by public or private institutions, this is registered as an increase in the productive output of the economy. We may refer to this form as *horizontal growth*. One of the ideological qualities of the concept of GDP, as a measure of economic growth, is that it serves to perfectly obscure the difference between vertical and horizontal growth. In quantum physics, we find the so-called uncertainty principle, stating that it is impossible to simultaneously measure the position and the momentum of a particle. GDP creates a similar uncertainty principle in economics whereby it is impossible to measure simultaneously the efficiency and the expansion of the market economy.

Economic growth not only constitutes the difference between the present and the past. The second part of the quotation from Mankiw and Taylor's textbook at the beginning of this section invokes the image of rich versus poor countries, the United Kingdom and the United States versus Burundi. Being confronted with the difference between $41,950 and $620, the reader cannot help thinking about what it would be like to live in Europe or the United States on less than $2 per day. This is not even enough to buy one of those Big Macs that have come to stand as the gold standard for international comparisons of purchasing power. In other words, people in Burundi have to decide whether they want to eat half a Big Mac each day or one Big Mac every other day. Fries and Coke are luxuries that they can only dream about. Thank heaven for economic growth which has spared people in our part of the world from such hardship.

The outstanding quality of the textbook quote is in the way that it reveals the ideological nature of the notion of economic growth as expressed through increasing amounts of GDP in a very pure form. The text barely needs analysis in order to unveil its ideological nature, as the analysis is included in the text itself. The remark in the second parenthesis notes how the measures of income per person in different countries 'are calculated by the World Bank, and, as is customary in international income comparisons of this kind, they are given in US dollars'. To understand the true significance of this 'custom', we only need to read the remark more literally than it immediately appears. The US dollar is presented as being in itself merely a neutral medium for the measurement of the standard of living in different countries. At the same time, the United States itself appears at the top of the list of countries in which Burundi is found at the bottom. We can understand this through the notion of 'concrete universality' that Žižek picks up from Hegel: '[T]he paradox of the proper Hegelian notion of the Universal is that it is not the neutral frame of the multitude of particular contents, but inherently divisive, splitting up its particular content: the Universal always asserts itself in the guise of some particular content which claims to embody it directly, excluding all other content as merely particular.'[19]

The US dollar appears as the 'universal frame' that allows for comparison between particular countries, while at the same time the United States is one of these particular countries. The United States constitutes simultaneously form and content. In the words of Žižek, the United States is that particular content which claims to be the direct embodiment of the Universal itself. All other countries are merely particular derivatives of this Universal. We can even proceed one step further to suggest that the United States is the sublime object of capitalist macroeconomics. Economic growth rates measure the speed at which particular countries are approaching this sublime state where form and content coincide. GDP is not just a measure of the 'standard of living' in individual countries. GDP is a measure of the extent to which life in different countries conforms to a universal capitalist way of living. In brief, this way of life is one where all production and all exchange take place within the framework of a market where all transactions are mediated by money. The market itself thus becomes the 'standard of living'. The United States is the incarnation of a society where life and market coincide. The United States is the sublime point on the horizon at which the trajectory of economic growth points. GDP is the gold standard of living.

It is interesting to note how Mankiw's, listing of the GDP per person of different countries, is at best a highly selective representation of the wealth of different nations and at worst a downright misrepresentation. Cross-checking the figures against data from the World Bank, we find it is true that the GDP per person in 2005 in the United States was approximately US$42,000.[20] However, this does not put the United States in first but rather only in eighth place in the world. In first and second place, we find Qatar and Luxembourg, with average personal incomes of US$69,500 and US$68,290 respectively. In sixth place, we find Norway, with an average GDP per person of US$47,620, rather than the US$40,420 of Mankiw's table.

A story goes that Charlie Chaplin once entered a Charlie Chaplin lookalike contest and only finished in third place. The point of this story, which may or may not be true, is of course that the outcome is an embarrassment for the contest itself. When Mankiw goes out of his way to construct a list of countries with the United States at the top, the function seems to be to save the very competition underlying the list. The fact that Qatar and Luxembourg are in first and second place is an embarrassment to the very concept of GDP as it is too obvious how their wealth is based on either unsustainable exploitation of finite natural resources or parasitic extraction of money from other countries by means of banking and financial services benefiting from bank secrecy laws and capital-friendly tax regulations. The high ranking of both of these countries undermines the image of GDP as a measure of the productivity and wealth of a nation.

Of course, economists are not idiots. Since the very inception of the measure of GDP, it has been well recognized that GDP is only a crude measure of the wealth and well-being of the people of a country. The father of GDP himself, Simon Kuznets, is famously quoted as saying that '[t]he welfare of a nation can scarcely be inferred from a measurement of national income'.[21] One might have expected such insight to lead to caution when using the measure of GDP in economic analyses and even greater caution when using it in politics. However, the function of Kuznets's reservation in contemporary politics seems to be to serve merely as a component in the ideological mechanism that Žižek refers to as cynicism: '[I]deology's dominant mode of functioning [today] is cynical.... The cynical subject is quite aware of the distance between the ideological mask and the social reality, but he none the less still insists upon the mask. The formula... would then be: "they know very well what they are doing, but, still, they are doing it."'[22]

As we apply this formula to the political economy of contemporary capitalism, we get the following: economists and politicians *know very well* that the welfare of a nation can scarcely be inferred from a measurement of national income *but still* they act in a way that is increasingly fixated on the optimization of economic growth as measured by GDP in their particular country, region or city. In order to understand why GDP is still the preferred measure of societal progress in contemporary economic and political planning, while most people agree that it is at best an inaccurate and at worst a misleading measure of societal well-being, we need to recognize that the primary purpose of the GDP measure is not to measure societal well-being but rather to promote certain conceptions of what an economy is in the first place. And for that purpose, the inaccuracy of GDP is a strength rather than a weakness. While GDP may be ill-suited for its original purpose of measurement, it is perfectly suited for its contemporary ideological purpose.

Real GDP

We have already seen how Žižek defines the 'function of ideology' as being 'not to offer us some point of escape from our reality but to offer us the social reality itself as an escape from some traumatic, real kernel'.[23] In the current context, this obviously begs the question: what is the 'insupportable, real, impossible kernel' that GDP serves to mask? Perhaps we can find an answer to this question simply by reading economic statistics literally. When developments in GDP are tracked over the course of a period of time, economic statistics typically invoke the concept of 'real GDP'. Rather than recording the value of an annual production of goods and services at their historical market prices, real GDP is calculated by applying the price levels of a particular year to the volume of output across a series of years, thus generating an index tracking changes in GDP volumes adjusted for inflation or deflation. The idea of real GDP is to separate changes in GDP due to changes in output volume from changes merely due to price fluctuations and inflation. Real GDP is distinct from nominal GDP that is not adjusted for inflation.

A similar form of reasoning is used in the comparison of GDP across different countries with different currencies. Rather than simply converting different national GDP measures according to currency exchange rates in the market, economic statistics invoke the concept of purchasing power parity (PPP). The idea is that the purchasing power of money in different countries varies in a way that is not reflected in the exchange rates in the market. For instance, 100 Danish kroner converts into about 240 Mexican pesos in the FOREX market, but if you spend 240 pesos on housing, food, energy and other life necessities in Mexico, you get a lot more goods than if you spend 100 kroner in Denmark on the same things. Economic statistics thus use an imaginary 'basket of goods' as a kind of gold standard to arrive at a 'real exchange rate' between different currencies reflecting their different local purchasing power. Instead of the immediate nominal exchange rate, the real exchange rate is used to compare the levels of GDP in different countries according to their value in terms of purchasing power.

While the concepts of *real* GDP and *real* exchange rate are indeed useful measures in many economic analyses, they are hardly consistent with Žižek's definition of the concept of the real. In the calculation of real GDP, we are merely generalizing the price levels of a particular year in the symbolization of the value of output over a series of years. However, this does not bring us any closer to the *real* value of the output produced. The price level of goods and services in the base year of the index is still nothing but a nominal expression of the value of these goods and services. In similar fashion, the concept of real exchange rate is ultimately a tautology that does not bring us outside of the domain of nominal prices. A genuinely Lacanian concept of real GDP would first of all insist that the value of the productive output 'resists symbolization'. This point is in line with classical economics in general, and Marx in particular, who maintains the distinction between exchange-value and use-value and insists that the latter cannot be reduced to the former. Secondly, and perhaps more importantly, such a concept would also point to the surplus of production and consumption in society that are not subject to pricing and exchange for money in the market. These are all aspects of natural capital and labour as a gift that we have been looking at in the preceding analysis.

Another dimension of real GDP lies in the relation between 'knowledge or technology' and money. Even though the measurement of the volume of GDP, capital and labour within an economic entity such as a nation-state is anything but unproblematic, it is still possible to come up with some kind of figure that appears to make sense. But how is it possible to quantify and express the aggregate volume of 'knowledge or technology' in a country? Should we measure the accumulated number of years that children and young people spend in schools and universities? But then what about the learning that takes place as people are actually working? And where do we draw the line between the learning of basic skills, such as walking, talking and using tools, that are required for human beings to even qualify as labour, and the more advanced capacities that qualify as 'knowledge or technology'? Could we then use the amount of money spent in the R&D departments of the companies of the country as a proxy for the level of 'knowledge or technology'? But then what about the ideas that people come up with spontaneously as they are busy doing other things? How do we even price the value of one year of education or one good idea in order to find a figure to put into the equation? And how do we price the value of the 'knowledge or technology' inherent in the common symbolic infrastructures of language, culture, tradition, history and so forth. The solution to these problems of measurement and pricing is much simpler than these questions might indicate. Instead of treating 'knowledge or technology' as a variable that is measured through the observation of empirical data and then entered into the right-hand side of the equation in order to calculate the output on the left-hand side, T functions as a residual category that ultimately serves to explain all those variations that cannot otherwise be explained through variations in the other input variables.

Barro and Sala-i-Martin invoke the following example to explain their above definition of 'knowledge or technology':

> Technology can improve over time – for example, the same amount of capital and labor yields a larger quantity of output in 2000 than in 1900 because the technology employed in 2000 is superior. Technology can also differ across countries – for example, the same amount of capital and labor yields a larger quantity of output in Japan than in Zambia because the technology available in Japan is better. The important distinctive characteristic of knowledge is that it is a *nonrival good*: two or more producers can use the same formula at the same time.... Hence, two producers that each want to produce Y units of output will each have to use a different set of machines and workers, but they can use the same formula. This property of nonrivalry turns out to have important implications for the interactions between technology and economic growth.[24]

Although this may seem like a very plausible explanation for the different levels of GDP in Japan and in Zambia, the statement is ultimately nothing but a tautology. We might ask the follow-up question: How do we know that 'the technology available in Japan is better' than in Zambia? And the answer is of course because 'the same amount of capital and labour yields a larger quantity of output in Japan than in Zambia'. Since we cannot think of capital that does not always already incorporate some kind of technology, and we cannot think of labour that does not always already incorporate some level of knowledge, the difference between Japan and Zambia is not simply a matter of a difference in their respective levels of technology and knowledge. In turn, the input variable T seems to point towards some mysterious kind of *surplus* 'knowledge or technology' over and above the kinds of basic knowledge and technology that are already incorporated into labour and capital. We have already explored knowledge and technology in light of Žižek's notion of *objet petit a* and we have seen how it seems to stand for some unknown X-factor, for what is in capital and labour more than capital and labour?

If we look further into the example of Japan and Zambia given in the definition of 'knowledge or technology', we may note that a major portion of the GDP of Zambia is constituted by the 'production' of copper. The economy of Zambia is largely based on the exploitation of natural resources, which are subsequently exported in a more or less unprocessed form. By comparison, the economy of Japan is much more diverse and it contains a much larger proportion of technologically advanced manufactured goods, such as electronic equipment, motor vehicles and chemicals. In very general terms, Zambia is an exporter of copper and other natural resources and an importer of manufactured goods, while Japan is an importer of natural resources and an exporter of manufactured goods. It is of course true that a key reason for this division of labour between Japan and Zambia is the availability of much more sophisticated machinery and highly skilled labour in Japan compared to Zambia. But in so far as output is measured in price, the difference in the amount of output between the two countries is also largely a function of the pricing of the commodities produced in each of the countries. If the price of copper increases relative to the price of manufactured goods, the GDP of Zambia also increases relative to Japan, even if the volume of copper, as well as the input of capital, labour and 'knowledge or technology', remains constant.

The point is that money or price are not independent variables in the production function. Since the variable T, 'knowledge or technology', works as a residual category that explains variations in output that cannot otherwise be explained through variations in any of the other two variables, it implicitly covers up the variations between the output of different countries, which are due to the way that this output is priced in the market. When we look at the difference between Japan and Zambia, it seems that the difference between the levels of output in the two countries has nothing to do with the fact that the value of the currency of Zambia has been systematically depressed because the country has been weighed down with growing amounts of debt that must be repaid in foreign currency. The economic conditions of Zambia are comparable to many other countries in the South that are also suffering from an overhang of large amounts of debt denominated in foreign currencies, first and foremost US dollars. As these debts grow due to compound interest, a situation of financial asymmetry between the 'soft currency' countries of the South and the 'hard currency' countries of the North is created. This means that the productive output of countries in the South, such as Zambia, which typically consists of raw materials, may be purchased in foreign currency at a price that is cheap relative to the price of the manufactured goods produced in the North. The debtor–creditor relationship between the South and the North means that there is always a surplus demand for foreign currency in the South required to make interest and debt repayments to the North. The North can take advantage of this as it is able to purchase commodities in the South using its own currency, which is then partially recycled to the North as interest and debt repayments. This is of course especially the case with the United States, as much of the foreign debt in developing countries is denominated in US dollars.

Of course, the macroeconomic mechanisms at work in the trade between 'developed' and 'developing' countries are much more complex than this brief and very general sketch. The philosophical point that I want to make is that the fuzzy variable 'knowledge or technology' effectively veils the ontological gap between the real of value and the symbolic expression of price. At the heart of neo-classical economics lies the idea of market equilibrium, which means that the market price of a commodity is always already a correct expression of the value of this commodity. It makes no sense within the neo-classical paradigm to speak of value in terms of anything but market price. In this sense, Wittgenstein's dictum, 'Whereof one cannot *speak*, thereof one must be silent',[25] seems to be a perfect account of the neo-classical approach to value. The Marxist idea of exploitation, which is the appropriation of the difference between the exchange-value (price) and the use-value (value) of labour by capital, is thus inconceivable within neo-classical economics. There is nothing but price, and the value of things is nothing but their price in the market. This also means that any form of financial exploitation, which destabilizes some currencies through compound interest on foreign debt and thus skews international exchange relations in favour of hard currency countries, is inconceivable within the framework of the production function. The value of a country's output is the price at which this output is sold, and this price is always already correct.

Without dismissing the fact that 'knowledge or technology' is indeed an important factor in the constitution of the production apparatus of a country, the point is that this variable has also the ideological function of being 'an "illusion" which structures our effective, real social relations and thereby masks some insupportable, real, impossible kernel'. The symbolization enabled by money makes it possible to compare the price of commodities that are qualitatively different, such as a tonne of copper and a Honda Civic. The symbolic expression of the real value of commodities in prices structures the effective exchange of these commodities according to the law of equivalent exchange. But in this process of symbolization, the 'insupportable, real, impossible kernel' of money itself is masked. When using the notion of GDP to compare the amount of productive output between different countries, we count how many commodities and services are produced in each of the countries. However, we fail to see where the money which is used to perform the trading through which the price of the commodities and services is derived is produced. The reason why the GDP of developing countries in Africa is low compared to developed nations such as Germany, France and the United States is not only that the level of 'knowledge or technology' in African countries is low but also that the money used in transactions between African and western countries is denominated in euros or dollars and ultimately produced by European or American banks, whether they be central or private.

In fact, we might even argue that the global monetary-financial complex that is made up by the nexus between central, commercial and investment banks, as well as international institutions such as the IMF, the WTO, the World Bank and the Bank of International Settlements, constitutes the pre-eminent example of 'knowledge or technology' determining the level of output and growth in different countries. This point may be demonstrated through a rewriting of Barro and Sala-i-Martin's definition of the variable T in the production function quoted above:

> The output produced by workers and machines cannot be priced in the market without a *formula* or *blueprint* that shows how they can be exchanged for money. This blueprint is what we call *financial knowledge or monetary technology*. Monetary technology can improve over time – for example, a larger share of the value produced by capital and labour may be appropriated by the banking and financial sector in 2000 than in 1900 because the monetary technology employed in 2000 is superior. The way that monetary technology interacts with the productive economy can also differ across countries – for example, the productive output of the same amount of capital and labour yields a much higher price in Japan than in Zambia because the global monetary technology favours Japan. The important distinctive characteristic of financial knowledge is that it is a *non-rival good*: several countries may be subject to the same formula of financial ideology at the same time.... Hence, two producers that each want to produce Y units of output will each have to use a different set of machines and workers, but their products are priced by the same financial formula in the world market. This property of non-rivalry turns out to have important implications for economic growth in different parts of the world.

When looking at the global distribution of money and resources in the world, it is difficult to avoid asking how such economic inequality between different countries and different regions is possible. The notion of 'knowledge or technology' provides a politically very convenient answer to this question: the global disparities in the wealth of nations are not due to structural imbalances in the monetary-financial complex of the world but rather to the simple fact that white people, as well as some yellow people, are smarter and more technologically advanced than black, brown and red people. This way of looking at growth may be exemplified by the following passage from Mankiw, Phelps and Romer's review of growth theory, where they pose the following questions:

> Average incomes in the world's richest countries are more than ten times as high as in the world's poorest countries. It is apparent to anyone who travels the world that these large differences in income lead to large differences in the quality of life. Less apparent are the reasons for these differences. What is it about the United States, Japan, and Germany that makes these countries so much richer than India, Indonesia, and Nigeria? How can the rich countries be sure to maintain their high standard of living? What can the poor countries do to join the club?[26]

While these questions may immediately seem to be guided by a benign interest in helping developing countries out of poverty, the way that they are posed effectively rules out the possibility that the wealth of rich countries is dependent on the poverty of poor countries; that the relatively high standards of living in the United States, Japan and Germany are somehow based on the ability to buy cheap labour, manufactured goods, oil, gas and other natural resources from India, Indonesia and Nigeria. Instead, the notion of GDP growth through increase in 'knowledge or technology' masks the zero-sum game elements of the global economy, thus creating the fantasy that all countries of the world may simultaneously 'join the club'.

Part III

'Economy or Ecology? Yes, Please!'

6

The Need to Grow

Eco-Analysis and Green Growth

Denmark and Danish companies are, and should continue to be, a driving force in the work to strengthen business-driven corporate social responsibility. Corporate social responsibility makes it possible to combine innovation, productivity and growth with social responsibility, sustainability and respect for human rights. /.../ It is the Government's aim that we jointly succeed in putting the financial crisis behind us and create a basis for new responsible growth and employment. The crisis has increased the need to focus on social responsibility and responsible conduct in order to face challenges in the form of climate change, scarcity of natural resources and human rights violations by companies, investors, consumers as well as the public sector.[1]

The old rules may say we can't protect our environment and promote economic growth at the same time, but in America, we've always used new technologies – we've used science; we've used research and development and discovery to make the old rules obsolete. /.../ It's not an either/or; it's a both/and. /.../ A low-carbon, clean energy economy can be an engine of growth for decades to come. And I want America to build that engine.[2]

[T]he world faces twin challenges: expanding economic opportunities for a growing global population; and addressing environmental pressures that, if left unaddressed, could undermine our ability to seize these opportunities. Green growth is where these two challenges meet and about exploiting the opportunities which lie within. It is about fostering economic growth and development while ensuring that natural assets continue to provide the resources and environmental services on which our well-being relies. It is also about fostering investment and innovation which will underpin sustained growth and give rise to new economic opportunities.[3]

These are quotes from two policy papers, by the Danish government and the Organisation for Economic Co-operation and Development (OECD) respectively, as well as from a speech by President Obama on climate change. Judging from these statements, global policy makers seem to have finally woken up to the challenges of climate change, resource depletion and ecosystem erosion. Not only does the ecological crisis call for immediate action but the transition to a more sustainable economy also holds the key to solving the economic crisis of unemployment, public deficits, rising levels of debt and absence of growth. The quotes are representative of a much wider spectrum of governments, organizations and corporations that are making the business case for sustainability.[4] As the popular environmental movements emerged in the 1960s and 1970s, sustainability was largely a radical utopian cause on the political left with the ultimate aim of fundamentally changing the system. Today, sustainability has found its way into the mainstream. As the radical critique has been adopted by contemporary ideology, sustainability seems to have become *the new spirit of capitalism*.[5] Business leaders, as well as heads of governments, have realized that sustainability can be commensurable with profitability at the individual company level and growth at the national level. 'Responsible growth', 'green growth', 'green capitalism', 'sustainable capitalism', 'sustainable growth' or 'responsible capitalism' are different neologisms that fuse apparent contradictions into one seemingly coherent concept thus pointing towards a future where it is possible to both have the cake and eat it. The above quotes are of course representations of the third position in the Lenin joke discussed in the introduction. Spelling out Obama, 'it's not an *either* ecology *or* economy; it's a *both* ecology *and* economy.' This was also the option provided by Willy:s supermarket: buying cheap organic bananas, you do not have to make the choice between ecology and economy. You can get both at the same time.

There exists already a large body of literature that is critical of these notions of 'sustainable growth', 'green capitalism' and similar third-way concepts.[6] While most commentators will agree that at least 'green' capitalism is better than 'brown' capitalism, views differ with regard to the questions of whether a 'greening' of capitalism is sufficient to curb or even reverse the ongoing trends in climate change, resource depletion, ecosystem degradation and so on. Some critics argue that the concept of green capitalism is nothing but the 'greenwashing' of unsustainable business practices.[7] Green capitalism is nothing but ideology in the good old-fashioned meaning of the word, where it serves to conceal the true state of affairs. Other critics argue that, even if the aspirations of 'greening' capitalism are indeed sincere, such efforts will only suffice if this greening includes a radical break with the underlying drive for perpetual growth in our existing form of economic organization. While some degree of 'relative decoupling' of economic growth from increase in material throughput is possible, the kind of 'absolute decoupling' required to prevent humanity from violating the ecological limits of the planet's carrying capacity is, if not theoretically then at least practically, impossible.[8] In order to stay within the so-called 'planetary boundaries' for a 'safe operating space for humanity',[9] global society (if such an entity exists) must reorganize in a way where the economy can function without constantly expanding.

At this stage in this analysis, it is again crucial to stay on the narrow path of eco-analysis. While there are indeed good and compelling reasons to agree with the critics arguing that 'green growth' or any of the other similar oxymorons do not provide adequate solutions to the contemporary ecological crisis and probably not even to the economic crisis, the purpose of eco-analysis is not to recount the logical arguments and comprehensive statistics backing the critique. '[T]he task of philosophy', according to Žižek, 'is not to provide answers, but to show the way we perceive a problem can be itself part of a problem, mystifying it instead of enabling us to solve it. There are not only wrong answers, there are also wrong questions. These wrong questions are what we call ideology.'[10]

If green growth is the answer, what is the question? Rearranging the very first sentence from the OECD paper, we can pose the following as the underlying question for which green growth is the answer: how is it possible to 'foster economic growth while assuring that natural assets continue to provide the resources and environmental services on which our well-being relies?'[11] The critique referenced above pointing out that green growth is just greenwashing, or in any case not sufficient to fully address the ecological crisis properly, does not immediately move beyond this question. Rather than posing green growth as the answer to the question, the critique argues that, green or not, economic growth is incompatible with natural assets continuing to provide the 'resources and environmental services on which our well-being relies'. However justified this critique, it still operates within the framework of the original question.

Contemporary books in the genre of critique of contemporary capitalism from the perspective of ecological economics typically consist of three elements: first, they list a number of facts and figures demonstrating how humanity is on a path to self-destruction. 'Our current lifestyle requires 1.6 planets to support itself.' 'In 2050, the world is going to run out of food.' 'The polar ice caps are melting at a higher rate than ever before.' And so on. Second, they demonstrate how the failure to take into account such issues are built into mainstream economic thinking as well as mainstream modes of production. Third, they describe a way out of this crisis by proposing new ideas for ways to reform the monetary system, the energy system, the political system, the means of production, the patterns of consumption or other dimensions of the contemporary economic system.

The point here is not to suggest that such books are in any way wrong or unimportant. The point is rather to maintain that eco-analysis should proceed in a different direction than such books, even if it shares many of their basic intuitions and concerns. There must be a division of labour between different disciplinary approaches to the same issue. Thus, the eco-analytical question is not whether economic growth and ecological sustainability are commensurable or not. This is the kind of 'wrong question' that Žižek is talking about. It may indeed be true that proponents of green growth provide a wrong answer to a wrong question. But from the perspective of eco-analysis, it is also true that critics within the school of ecological economics provide a right answer to a wrong question. With eco-analysis, we should instead be posing the following question: Why is economic growth so important in the first place? Why is it that even in the face of impending ecological catastrophes, contemporary capitalism still remains unconditionally committed to perpetual growth? What are the dynamics behind the seeming growth imperative and what is the function of economic growth? It would be misleading to suggest that discussions within ecological economics, as well as adjacent fields of inquiry, have not already touched upon these questions.[12] Still, the contention here is that eco-analysis may provide its own approach to the issue.

What We Don't Know that We Know about Climate Change

In order to explicate how the issue of economic growth shall be approached in this part of the book, we can compare the eco-analytical approach to the eco to the way that psychoanalysis approaches an analysand in therapy. Imagine that an obese woman comes into psychoanalytical therapy. She is suffering from excessive eating habits that are now threatening her physical health and perhaps even her life. Would we approach her by weighing her, measuring her blood pressure and submitting her to a number of other medical tests and then proceed to explain to her how a continuation of her current lifestyle gives her 80 per cent likelihood of dying within the next five years? And would we tell her that the cure for her problems is a healthy diet and daily exercise? The answer is obviously no. Psychoanalysis is anything but such a purely medical and physical approach to the analysand. Psychoanalysis approaches the analysand not merely as a body but as a subject. The strictly medical approach to a person suffering from obesity is, for a number of reasons, likely not only to be ineffective, it might even turn out to be counterproductive.

First of all, the obese analysand is probably already well aware that eating lots of sugar and fat, drinking lots of Coke and spending most of your time in front of a television set is not very healthy. Psychoanalysis would not regard the problem as lying on the level of this form of conscious knowledge. Instead, the psychoanalyst would perhaps try to uncover the function of food in the analysand's structure of desire. What does food represent for the analysand? How does the analysand perceive or even imagine her own body? Which kinds of emotional conflict does the eating function relieve for the analysand? And so on. The analysand may even be fully aware of the damage she is doing to her own body but this damage could be the very purpose of her eating behaviour. Second, even if some of the information about the relations between diet, exercise and her physical condition is new to the analysand, psychoanalysis does not expect her to be able immediately to convert it into new patterns of behaviour in relation to eating, exercising and other bodily practices. If the analysand is perhaps already alienated in her relation to her own body, the objectifying knowledge of kilograms, BMI, calories, blood pressure counts and so on may serve merely to perpetuate this feeling of alienation, thus triggering even further excessive eating. In the terminology of Rumsfeldian epistemology, which we encountered in a previous chapter, the psychoanalytic solution to obesity is not merely to increase what the analysand *knows that she knows* about diets, nutrition, exercise and the physical state of her own body. The psychoanalytic solution is to work on the things that the analysand *doesn't know that she knows* about food, eating, her body and so on. As always in psychoanalysis, the solution lies in the unconscious.

Of course, there are major differences between an individual obese person, working to change dysfunctional habits of eating, and a global conglomerate of political institutions, private corporations, individual citizens, NGOs and so on, working to change dysfunctional systems of producing, consuming and distributing goods and services. Still, the analogy serves to help raise the question of whether the way that data and scientific knowledge about climate change, natural resource depletion, ecosystem destruction and so on are reported and used in campaigns to create societal change is not just part of the solution but perhaps even also part of the problem. The Club of Rome's report on *The Limits to Growth* in 1972 was the first in a long series of reports and papers documenting the detrimental ecological effects of an expanding capitalist economy and predicting very dire outcomes for 'mankind' unless current trends were reversed. A more recent report in the same genre was published in 2014 by the American Association for the Advancement of Science (AAAS) titled: *What We Know: The Reality, Risks and Response to Climate Change*. This is the main message of the report:

> 'The overwhelming evidence of human-caused climate change documents both current impacts with significant costs and extraordinary future risks to society and natural systems. /.../'
>
> Surveys show that many Americans think climate change is still a topic of significant scientific disagreement. Thus, it is important and increasingly urgent for the public to know there is now a high degree of agreement among climate scientists that human-caused climate change is real. /.../
>
> It is not the purpose of this paper to explain why this disconnect between scientific knowledge and public perception has occurred.... Instead, we present key messages for every American about climate change:
>
> 1. Climate scientists agree: climate change is happening here and now. /.../
> 2. We are at risk of pushing our climate system toward abrupt, unpredictable, and potentially irreversible changes with highly damaging impacts. /.../
> 3. The sooner we act, the lower the risk and cost. And there is much we can do.[13]

The point here is not to contest the scientific validity of the propositions made by AAAS as well as the other 97 per cent of all climate experts agreeing that human-caused climate change is happening. The point is also not to downplay the severity of the risks stressed by the AAAS as well as the other climate experts. Instead, the point is to suggest that there is room and a need for other kinds of analyses that do not get caught up in the rhetoric of objectifying scientific knowledge. Scientific studies may unveil causal relations between industrial forms of production and the disturbance of the balance of natural ecosystems, and they may project these relations into the future, showing their ultimate catastrophic effects. The problem, which we explored in part I, is that these studies lack a theory of humanity explaining why it is at all worth saving in the first place. Humanity itself is not included in the view of the world of the natural sciences. They tend to conceive of the world in purely physical terms, thus excluding the domain of metaphysics. In the words of Peter Sloterdijk: 'Humans are beings that participate in spaces unknown to physics.'[14] The world is not merely a conglomerate of different ecosystems. It is also the place where human beings reside. This is the duality I have tried to capture with the concept of the eco.

The AAAS report makes a very insightful comment: 'It is not the purpose of this paper to explain why this disconnect between scientific knowledge and public perception has occurred.' This comment has almost the same structure as the famous liar's paradox stating: 'This sentence is false.' Inadvertently, the AAAS comment itself provides the answer to the question, which it does not intend to explain. The reason why scientific knowledge and public perception have become disconnected is precisely because scientific knowledge fails to account for the disconnect between scientific knowledge and public perception. When 'the public' is sceptical about the kind of scientific knowledge demonstrating human-caused climate change, it is perhaps wrong to conceive of this as a discrepancy between two forms of knowledge that are on an epistemologically equal level. While natural science can afford to describe the world in purely objectivist terms, 'the public' must develop a conception of the world that includes a meaning and purpose with its own being-in-the-world. As we have already discussed in part I, the idea of history moving towards major destructive catastrophes is largely incompatible with the life-world experience of a human subject. How can I possibly make sense of such a condition of existence? The report even provides some empirical basis for this claim as it states that 'while the public is becoming aware that climate change is increasing the likelihood of certain local disasters, many people do not yet understand that there is a small, but real chance of abrupt, unpredictable and potentially irreversible changes with highly damaging impacts on people in the United States and around the world'.[15]

The truth of this statement is perhaps more profound than was initially intended. The reason why the public is only able to conceive of 'certain local disasters' but not 'abrupt, unpredictable and potentially irreversible changes' is not because the public is stupid and misinformed (they may be that, too, though) but because the chance of such changes is indeed 'real' in the true Lacanian sense of the word. Such catastrophic changes 'resist symbolization' in the sense that they cannot be integrated into the life-world experience of an individual subject. I would risk the claim that every day, in order to preserve their sanity, even climate scientists themselves have temporarily to deny the prospects of catastrophic climate change when they go home from work to spend time with their children. The potential of eco-analysis in this context is that it allows us to acknowledge the refusal of 'the public' to incorporate the scientific knowledge of climate change and other catastrophic events without denying the scientific validity of this knowledge. In this sense, the AAAS disclaimer is an implicit statement of a key eco-analytical question: 'It is the purpose of eco-analysis to explain why this disconnect between scientific knowledge and public perception about climate change has occurred.'

Let us summarize this section by completing the analogy with obesity. The risk of subjecting the obese woman to a purely medical and physical gaze is that it might inadvertently confirm her own unconscious knowledge that her body is indeed just a heap of flab and meat that is heading for complete self-destruction and not worthy of any kind of edifying care and tenderness beyond the quick stimuli of sugar, fat and salt. And also her subjective sense of self-confidence and worthiness may be shattered by the realization of the things she has done to her own body. The calls for global action against climate change based on new and evermore gloomy scientific projections risk functioning in a similar fashion. Why care about nature if it is broken anyway? And why try to save the human race when our biggest accomplishment so far has been to destroy our own means of existence on the planet? Rather than repeating medical facts about obesity, health and nutrition, psychoanalysis would approach the obese woman by trying to uncover the way that her desire is structured around certain unconscious ideas about food, body and eating. In similar fashion, eco-analysis should not simply repeat the scientific facts about climate change, resource depletion and so on but rather attempt to uncover why contemporary capitalism seems to be attached to growth. Green growth is the equivalent of trying to fight obesity by eating fat-free potato chips, drinking Coke Zero and watching television shows about other people who are trying to lose weight.

From the perspective of eco-analysis, the problem with green growth is not that it is impossible or insufficient in terms of solving the ecological crisis. It is rather that the notion of green growth comes to stand in the way of an inquiry into the notion of growth itself. As long as the obese woman believes that she can solve her problems by eating fake fat and fake sugar instead of the real thing, she is not confronted with the core existential question: Why must I keep eating so much? In similar fashion, the fantasy of green growth stands in the way of a thorough *Auseinandersetzung* with the key question of the eco: Why must the economy keep growing? The purpose of the following analyses is to move beyond some of the implicit assumptions inherent in the notion of green growth in order to engage with this question of the apparent imperative of growth.

Monsanto with Brundtland

Another euphemism for the notion of green capitalism, which aims to reconcile ecological sustainability with continued economic growth, is the concept of 'sustainable development'. As we have already touched upon in part I, this concept is presented in the so-called Brundtland Report on *Our Common Future*. Perhaps the most quoted passage from the report is exactly the definition of sustainable development, which reads like this:

> Sustainable development is development that meets the needs of the present without compromising the ability of future generations to meet their own needs. It contains within it two key concepts:
>
> - the concept of 'needs', in particular the essential needs of the world's poor, to which overriding priority should be given; and
> - the idea of limitations imposed by the state of technology and social organization on the environment's ability to meet present and future needs.[16]

Through the concept of 'essential needs', this definition invokes an understanding of economic growth, which is comparable to the one found in physiocrat economics. For the physiocrats, the function of the economy is the provision of goods to satisfy very basic human needs. The conception of value as *blé* suggests that the ultimate purpose of the economy is to feed people. The reasoning behind the Brundtland definition of sustainable development is that if the economy grows in a way that is incompatible with the maintenance and reproduction of natural ecosystems in the environment, it ultimately erodes its own capacity to meet the needs of the people within the economy. If the consideration for the satisfaction of the needs of future generations is included in the organization of the economy today, the preservation of natural ecosystems is automatically implied in economic development. Still, according to the Brundtland Report, the economy must grow in order to meet the needs of present and future generations. The growth imperative is derived from a double concern for currently existing poor people, whose essential needs are not being met, but also for the future prospect of a growing population of people, whose needs must be met.

On the one hand, the Brundtland vision of sustainable development is sympathetic and undoubtedly well meaning. It is difficult to argue against concern for hungry people. On the other hand, the recourse to the 'essential needs' of 'present' and 'future generations' creates an ideological deadlock which makes it impossible to address the fundamental issue of growth. A further eco-analysis reveals that the attachment to 'the essential needs of the world's poor' is what keeps truly sustainable development in check. Support for this claim may be found by looking into the way that the multinational agricultural corporation, Monsanto, presents its business and purpose in the world.

Monsanto is leading the field in the production of genetically modified organisms (GMOs). Their business model combines the sale of seeds that have been genetically engineered to resist particular pesticides and herbicides and the sale of those very same pesticides and herbicides. Monsanto's most well-known product is the herbicide Roundup that corresponds to particular brands of seeds sold as 'Roundup Ready'. The business model also includes strong enforcement of biological patents, as well as restricting the rights of customers to reuse seeds from their harvest, thus requiring farmers to repurchase new seeds from Monsanto for every season. The company has adopted the following slogan: 'The Challenge: Meeting the needs of today while preserving the planet for tomorrow', and in 2008, Monsanto made a 'commitment to sustainable agriculture'. The company presents the implications of this commitment with the image in Figure 6.1 and the text below.

Our Commitment to Sustainable Agriculture
Producing More. Conserving More. Improving Lives.
Our vision for sustainable agriculture strives to meet the needs of a growing population, to protect and preserve this planet we all call home, and to help improve lives everywhere. /.../

Figure 6.1 Monsanto campaign image

Producing More
Monsanto works with farmers from around the world to make agriculture more productive and sustainable. Our technologies enable farmers to get more from every acre of farmland. /.../

Conserving More
We've strengthened our goal of double crop yields by committing to doing it with one-third fewer resources such as land, water, and energy per unit produced. /.../

Improving Lives
The technology we use to develop better seeds and the partnerships we nurture to develop new agronomic practices can drive big increases in yield and productivity. For all the world's farmers who raise themselves from poverty to prosperity, many more people will also prosper, through healthier diets, greater educational opportunities, and brighter futures fueled by more robust local economies.[17]

Taken at face value, Monsanto's formulation of 'the Challenge' seems to be perfectly aligned with the goal of sustainable development that was put forward in the Brundtland Report. And the image of an African mother with her daughter seems to indicate that Monsanto also shares the view of the report that 'overriding priority should be given' to 'the essential needs of the world's poor'. In this view, the Brundtland Report must be regarded as a success. The visions put forward in the report have trickled out into society and have now become incorporated into the business model of a major capitalist multinational corporation such as Monsanto that is a market leader in its own field. Furthermore, Monsanto has taken in the projections of future trends in world populations, which can also be found in the Brundtland Report. Today, it is being estimated the world population is going to increase by more than two billion people over the next three decades. The conflict between such projections and the need to restrain exploitation of natural resources in order to save our environment is resolved in Monsanto's double imperative of 'producing more' while at the same time 'conserving more'. This challenge is met by developing new seeds and new crops that enable farmers to increase yields not only in absolute terms but also relative to the amount of land, water and other natural resources used in the process.

Monsanto is not only the biggest agrichemical company in the world. Through a long list of environmentally unfriendly business practices around the globe,[18] Monsanto has earned the position as enemy number one in those parts of the environmental movement that are concerned with food and agriculture. In this light, it is easy to write off Monsanto's mission statements and their commitment to sustainable agriculture as nothing but a smokescreen for a business model that is anything but sustainable. While such a critique is both true and reasonable, the discrepancy between words and action is perhaps not the most central point in our present context. What makes Monsanto interesting for my analysis is the way that the concern for present as well as for future generations of poor people in undeveloped regions of the world creates an alignment between environmental sustainability on the one hand and economic growth, efficiency and productivity on the other. In this way, the Brundtland Report provides perfect justification for the continued growth of Monsanto. Ecology and economy become two sides of the same coin.

As we have seen in part I, the discourse of ecology renders Man an ambivalent form of Being that is both inside and outside of nature. Man is 'included out' in the discourse of ecology. When concerns for a growing population of poor people in undeveloped parts of the world is introduced into the discourse about sustainability and the 'essential needs of the world's poor' are given 'overriding priority', it is almost as if these people are simply 'included in' on the side of nature. When it comes to overweight Americans, their consumption patterns may be deconstructed and dismissed as largely the result of unnecessary desires. They have very little moral value in terms of justifying the need for increasing outputs in global agriculture. However, the acute hunger of billions of future babies in Africa and Asia appears to be beyond any form of critical deconstruction or moral relativism. Their hunger constitutes 'essential needs' rather than constructed desires. The basic need for food is one of the things that human beings share with animals. It is part of what makes us natural beings. This is why the reference to 'essential needs of the world's poor' is such an efficient way of explaining and legitimizing perpetual economic growth.

The problem with the recourse to 'essential needs' in the order of the real is not merely that it provides legitimacy for a continuation of our current economic paradigm of perpetual growth. Closer eco-analytical scrutiny soon reveals that the concept of needs itself is highly elusive. This is a theme that I have already touched upon in the discussion of physiocrat economics. In order to explore this theme, I shall make a small theoretical excursus into the psychoanalytical distinction between need and desire. Here is how Lacan conceives of the relation between the two: 'Desire begins to take shape in the margin in which demand rips away from need, this margin being the one that demand – whose appeal can be unconditional only with respect to the Other – opens up in the guise of the possible gap need may give rise to here, because it has no universal satisfaction (this is called "anxiety").'[19]

The human being is born into the world with a series of fundamental needs that must be met in order for it to survive. These are needs such as hunger, thirst, sleep and shelter. In this series of needs we also find, however, the need for these needs to be met by the Other. This is what sets the whole process of subjectivation in motion. We could refer to this as the need for love. When a baby stops crying as it is given milk from the breast, it is not merely because its thirst is satisfied; it is also because it experiences how this need is met by the mOther. In turn, this means that the milk and the breast are not only objects that satisfy the baby's hunger and thirst. They also function as tokens of the love of the mOther. When the baby expresses a demand for milk by crying, this is not merely a 1:1 representation of the need for food. The demand over-determines the need for food in so far as it also contains a demand for love. This is what Lacan is referring to in the above quote. The representation of the need through the demand is always already caught up in the logic of symbolization that was discussed earlier. It is in the space between the need and the demand for its satisfaction that desire emerges. While the need for food may be fully satisfied, the need for love is ultimately insatiable. How does the baby know that the mOther feeds it because she loves it and not just because it is hungry? There is no object that may serve as an ultimate token of love beyond any doubt. The demand for love has 'no universal satisfaction' which is why it is correlative to 'anxiety'. The need for love thus opens up the space where phantasmatic projections create objects of desire. The need for milk becomes a desire for an object that also serves to confirm the love of the Other. The insatiability of the need for love is what propels the process whereby the person is integrated into the symbolic order of language, law and signification. As we have touched upon previously, this process is structured as a catch-22 of symbolic castration. On the one hand, the appropriation of language provides the person with the means to more accurately express his or her desires. On the other hand, this very symbolization also creates the surplus inherent in desires in a way that prevents full satisfaction.

When concerns for the 'essential needs of the world's poor' are used as explanation and justification for the necessity of economic growth, the intricate relation between need, demand and desire is overlooked. Between 1800 and 2011, the world population grew from one to seven billion. Furthermore, the world population is projected to exceed nine billion around 2050. It is of course true that this represents a growth in the sheer volume of basic human needs that require satisfaction, the amount of mouths that need to be fed.[20] But just as the infant must express its needs in a way that is recognized by the mOther in order to be satisfied, so must the needs of a growing world population be expressed in a way that is recognized by the Other of the Market in order to result in economic growth. The way that needs are expressed to be recognized by the market is of course in the form of demand. Demand is, however, not a simple 1:1 representation of needs.

First of all, people must have money in order to have their needs recognized as demands by the market. While there are indeed many hungry people in poor parts of the world, this does not necessarily translate into a demand for food. We have previously seen how the market works according to the law of equivalent exchange. Needs must be able to conform to this law before they can be expressed as demands. Or perhaps, rather, it is the conformity to the law of equivalent exchange that turns needs into demands. If people do not have something to offer in exchange for the satisfaction of their needs, the needs elude the symbolic order of the market economy. The needs come up against the law of impossible exchange. This is why we cannot take for granted that economic growth is necessarily going to meet the needs of a growing population of people.

Second, the expression of a need as demand is inseparable from a transformation of the need itself. Just as the infant is not merely satisfied to have its material needs met but also demands to become the object of love, economic subjects are also not satisfied just to have their material needs met by the market. They also demand to become objects of the demands of the market. When the infant demands love, it demands to become the object of the demands of the mOther. This is the point where 'demand rips away from need' and 'desire begins to take shape'. We can transpose this point onto the domain of eco-analysis by paraphrasing the quotation from Lacan above: 'Economy begins to take shape in the margin in which demand rips away from need.' It is at that moment that demand exceeds basic material needs that the order of the economy emerges. Or perhaps the other way around: as the order of the economy emerges, demand exceeds basic material needs. The integration of the subject into the symbolic order of the market through the expression of his needs as demands is also the shaping of the subject as a subject of desire. In this integration, needs become indistinguishable from desires.

We can elaborate this further through Lacan's statement that 'man's desire is the Other's desire'.[21] This means that the subject's desire is not merely the desire for other objects but also the subject's desire to become the object of the Other's desire. We have already touched upon this dialectic in the previous section on the role of money in economic castration. When subjects engage with the market, they not only express their needs through demands for the commodities that are traded in the market. The economic subject also demands to become the object of the demands of the market itself. While agents trading in the market express very different demands, the market itself is the incarnation of the demand for money as such. Through the mirroring of this demand, the subject becomes an economic subject. It is through the possession of money that the economic subject becomes the object of the desire of the market. When needs are expressed in the form of demands to the market, this demand always already includes a demand for money. The market is not merely the site where demands are matched with corresponding supplies. The market is also a supplier of demand itself. This is first and foremost expressed in the desire for money. Of course, the desire for money may be practically indistinguishable from the need for food in so far as money can be exchanged for food. This is comparable to the way the infant's desire for love may be practically indistinguishable from its need for milk in so far as breastfeeding is also a symbol of the love of the mOther. The Brundtland Report calls for the economy to meet 'the essential needs of the world's poor'. Implicit in this call is the idea that economic growth will do nothing but increase the volume of food and other basic commodities, which will then subsequently become available for distribution among these poor and hungry people. We see here the reminiscence of the physiocratic idea that economic growth amounts to an increase in the production of *blé*. But meeting these needs through economic growth invariably also leads to the transformation of these needs into desires. The essential needs of the world's poor are indistinguishable from their desire to no longer be poor.

We may push the argument even further to suggest that behind the apparent humanism of the Brundtland Report is a deep ethnocentric post-colonial racism. While the inclusion of first-world people in the global order of market economy has indeed sparked aspirations of wealth and consumption far beyond the satisfaction of basic essential needs, the inclusion of third-world people in this order through further economic growth is assumed to leave their primitive subjectivity intact. As their essential needs have been meet, they will happily stay at the bottom of the Maslowian pyramid without developing the same insatiable desires for luxury cars, weekend trips to Paris, the latest iPhone or even just more and more money sitting idle in a bank account that plague the people of the first world.

What Would Keyser Soeze Do?

In philosophical terms, the reference to 'the essential needs of the world's poor' which we find in the Brundtland Report, as well as Monsanto's commitment to 'meet the needs of a growing population', function to provide what Žižek calls an 'answer of the real'. Žižek defines this concept in relation to communication:

> For things to have meaning, this meaning must be confirmed by some contingent piece of the real that can be read as a 'sign'. The very word *sign*, in opposition to the arbitrary mark, pertains to the 'answer of the real': the 'sign' is given by the thing itself, it indicates that at least at a certain point, the abyss separating the real from the symbolic network has been crossed.[22]

It is important to note that the gap between the real and the symbolic is not bridged through a relation of representation. The answer of the real emerges in a space of indistinction between the real and the symbolic, thus establishing a so-called 'quilting point' [*point de capiton*] where the real is weaved into the fabric of reality. Such quilting points are in turn the precondition for all other relations of representation between the symbolic and the real. In the context of sustainability, 'the essential needs of the world's poor' provide an answer of the real to confirm that human beings are also part of the natural order and thus also a concern for ecology. It is as if human beings, through this reference, are included as one of those endangered species that must be saved as part of the conservation of our global biosphere. Furthermore, the increasing number of people on the planet also seems to be regarded as a fact of nature that has to be accommodated as part of sustainable development. The ideological result is the above-mentioned alignment between the Brundtland Report and the business of Monsanto. What happens is a conflation between the ecological and the economic concept of sustainability. More specifically, this means that economic growth becomes intertwined with ecologically sustainable development. We need to increase economic output in order to feed future generations of poor people. GMOs and pesticides are part of an ecologically sustainable future because only through such technological innovations is it possible to feed the people of the world while at the same time conserving natural resources. It may even be argued that more organic forms of farming are less environmentally friendly because they produce a smaller yield relative to the amount of land occupied. Organic farmers are irresponsible in so far as they do not produce as much food to feed the world's hungry as they could.

As much as I am not only critical but also petrified by the business practices of Monsanto and like-minded agrichemical corporations, it is beyond the scope of this book to make any qualified contribution to the existing research on the ecological effects and hazards of the introduction of pesticides and GMOs into natural environments. Instead, I shall proceed to think through a way of breaking out of the ideological deadlock inherent in the concept of sustainable development and the reference to the essential needs of the world's poor. Žižek provides a recipe for ideological intervention that can help us here:

> [I]n order effectively to liberate oneself from the grip of existing social reality, one should first renounce the transgressive fantasmatic supplement that attaches us to it. In what does this renunciation consist? In a series of recent (commercial) films, we find the same surprising radical gesture.... when, in the flashback scene from *The Usual Suspects*, the mysterious Keyser Soeze returns home and finds his wife and small daughter held at gunpoint by the members of a rival mob, he resorts to the radical gesture of shooting his wife and daughter themselves dead – this act enables him mercilessly to pursue members of the rival gang, their families, parents and friends, killing them all. /.../ This act, far from amounting to a case of impotent aggressivity turned against oneself, rather changes the co-ordinates of the situation in which the subject finds himself: by cutting himself loose from the

precious object through whose possession the enemy kept him in check, the subject gains the space of free action. Is not such a radical gesture of 'striking at oneself' constitutive of subjectivity as such?[23]

What would such 'striking at oneself' mean in the context of the Brundtland Report and in the context of green capitalism in general? And what is the 'precious object through whose possession' the well-meaning efforts to promote sustainable development seem to be kept in check? Perhaps what is needed in order to 'gain the space of free action' with regards to ecology and politics is to cut ourselves loose from the 'precious object' of present and future generations of poor people and especially from the idea that we need to increase output in order to feed a growing world population that is approaching nine billion people. The woman and the child in the image from the Monsanto presentation may serve as an illustration of this object that we would need to renounce. They constitute, in other words, the functional equivalent to Keyser Soeze's wife and child. We should no longer take as premise for the organization of our economy, our production apparatus and the way we grow our food that these things need to be designed in a way that is able to fulfil the needs of even more people than today.

The point of such a seemingly cruel, cynical and 'crazy' renunciation is, as noted, to disconnect political efforts to curb ecological degradation from an economic imperative of growth. What is so eminently illustrated in the Monsanto case is how the poor people of the world are taken hostage by corporate interests in the struggle to define sustainable development. Critiques of GMOs, pesticides and other technological innovations may be denounced as in solidarity with people in undeveloped parts of the world. This functions to paralyze critical positions on the traditional political left for whom solidarity with the poor is fundamental. Their aim to resist the paradigm of growth is neutralized by their own abstract humanism. Unless of course we believe that Monsanto has found the solution to the current ecological crisis, we need to break out of this deadlock. Keyser Soeze illustrates a radical way of doing this. Paraphrasing Žižek, we may formulate the situation in the following way:

> In order effectively to liberate oneself from the grip of existing social reality, where economic growth through increased productivity and technological innovations seems to have become an inherent imperative in most political positions across the traditional left–right spectrum, one should first renounce the transgressive phantasmatic supplement of the need to feed a growing world population that attaches us to this social reality.

The reason why we should renounce the need to feed future generations of poor people is not of course that these people do not deserve to be fed. Quite the contrary, it is very likely that our commitment to feeding them will have the completely opposite effect. As long as we stick to this commitment, we provide legitimacy for exploitative business practices by corporations such as Monsanto and Cargill, which have detrimental consequences not only for global ecosystems but also for the economic and social viability of local communities, where they operate. When trying to imagine our common future, we should not be posing the following question: How can we transform our societies towards ecologically sustainable lifestyles and production while at the same time building the capacity to feed a world population in excess of nine billion people? It is as if the second part of the question functions to annul the first part. Instead, we should rather just stick to the first part and merely ask: How can we transform our societies towards ecologically sustainable lifestyles and production? In this way, we may think of solutions that point beyond the current imperative of economic growth. Paradoxically, such solutions may end up providing the best solution also for the poor parts of the world anyway. Instead of multinational agrichemical corporations committing themselves to 'help' feed the poor people of the world, many of these people might be better off just being left alone in their efforts to feed themselves.

7

The Desire to Grow

Consumption and Economic Castration

> [F]antasy designates [the] unwritten framework which tells us how we are to understand the letter of the Law....[S]ometimes, at least – the truly subversive thing is not to disregard the explicit letter of Law on behalf of the underlying fantasies, but to stick to this letter against the fantasy which sustains it.[1]

If we were to take the concept of green growth literally, it might simply refer to the kind of organic growth of plants that we find everywhere in nature. Plants are green and they grow. But of course, as with any other ideological phrase, the first thing you need to know about green growth is not to take it literally. The 'growth' in 'green growth' still refers to good old-fashioned economic growth as it is measured by econometrics and expressed through annual figures of GDP. This means that the concept of green growth retains the narrow understanding of 'economy' that has been defined and delineated through the discipline of economics. Economy refers to the total of goods and services that are being exchanged for money in the market. And growth is about increasing the volume of such goods and services.

GDP is not only a measure of the volume of production of an economy. It is simultaneously a measure of the volume of consumption of an economy. When production grows, consumption also invariably grows. Only goods and services that are consumed count as being produced. It is true that a surplus of production over consumption may be carried over into subsequent cycles of production and consumption as investment, but ultimately the economy can only keep growing if both production and consumption increase. Green growth is therefore not only a call for a transformation and expansion in the current mode of production. It is also a call for an expansion in the current mode of consumption. From the perspective of eco-analysis, this is another interesting dimension of the notion of green growth. As the imperative of growth is retained in the vision of a 'green' future, fundamental questions of consumption are left unexamined. One such question is: Why must we keep consuming more?

In the previous chapter, we touched upon some of the mechanisms whereby the economy not only produces supplies to meet existing demands but is also itself a supplier of demands. This line of analysis continues in the current chapter. The creation and expansion of consumer demand for the goods and services produced and offered in the market are not only brought about through the conflation of needs and desires. They are also an effect of a particular paradox inherent in the division of labour.

On the one hand, division of labour constitutes an improvement in the productive capacity of society. The output of the economy is drastically increased and more commodities and services become available for consumption. In our exploration of Adam Smith, we have already seen how the relation between the division of labour and the increase in productivity is at the heart of the classical conception of the modern economy. The division of labour is the *condition of possibility for the economy's capacity to satisfy the wants of the consumer subjects*. On the other hand, division of labour is not only the division of labour from itself but also the division of labour from consumption. Division of labour is not only the organization of work into separate tasks, which increases the efficiency of labour. It is also the division of the act of production from the act of consumption. Rather than immediately consuming the fruits of his labour, the modern worker exchanges his labour for money, which is then again exchanged for consumer objects and services in the market. With eco-analysis, we can argue that the division of labour and consumption is at the same time the *condition of impossibility for the complete satisfaction of wants of the consumer subjects*. The mediation of the subject's desires through the economy is also a shaping of these desires whereby their full satisfaction is rendered impossible. This claim is the eco-analytical implication of Žižek's concept of symbolic castration that we have already encountered. The division of labour performs a form of economic castration. We shall expand on this concept by returning to the story of Adam's fall that was also discussed in part II.

The tree from which Adam picks the apple is the Tree of Knowledge. This suggests a contradiction between knowledge and the absolute *jouissance* found in the Garden of Eden. In similar fashion, the subject becomes subject by gaining knowledge about his own desires through the symbolic order. In the context of economy, the consumer gains knowledge about his own desires by venturing into the market offering different objects to match these desires. With this reflexivity, a kind of alienation is installed in the subject's self-relation which renders impossible the full complementation of the subject, makes impossible the absolute *jouissance*. '[A]ccess to knowledge is then paid with the loss of enjoyment – enjoyment, in its stupidity, is possible only on the basis of certain non-knowledge, ignorance' (Žižek 1989: 68). In other words, the price for becoming a subject is to deposit in the symbolic order a piece of enjoyment, a remainder, so that all future enjoyment can only become a derivative of the absolute enjoyment. This 'piece of enjoyment' has a rather complex status in the symbolic order. The story of Adam's fall illustrates how it was never there in the first place. Nevertheless, it now plays a crucial role in the symbolic order. We may say that enjoyment circles around within this order. It does not, however, circulate as real but as virtual. The image is projected into the symbolic order so that the subject may find here its lacking part, that the symbolic order does contain the possibility for absolute enjoyment. In this way, the subject is linked to the symbolic order.

Returning to the exposition of classical economy in part II, it is relevant that Adam Smith introduces the division between labour and consumption in the chapter about money. Smith provides his own classic account of the way that money emerges as a practical device to solve the inconveniences inherent in a moneyless barter economy. This is the story about the butcher, the baker and the brewer, who figure out how to use gold as a general medium of exchange, rather than bartering meat, bread and beer:

> [W]hen the division of labor first began to take place, this power of exchanging must frequently have been very much clogged and embarrassed in its operations.... The butcher has more meat in his shop than he himself can consume, and the brewer and the baker would each of them be willing to purchase a part of it. But they have nothing to offer in exchange, except the different productions of their respective trades, and the butcher is already provided with all the bread and beer which he has immediate occasion for. No change can in this case be made between them.... In order to avoid the inconvenience of such situations, every prudent man in every period of society, after the first establishment of the division of labor, must naturally have

endeavored to manage his affairs in such a manner, as to have at all times by him, besides the peculiar produce of his own industry, a certain quantity of some one commodity or other, such as he imagined few people would be likely to refuse in exchange for the produce of their industry.[2]

The account seems very intuitive: barter is the original form of economic exchange but, as the division of labour progresses, it becomes difficult and inefficient. This is when money emerges as a generalized medium of exchange to make the exchange of the different products of labour much easier. In philosophical terms, money is nothing but a particular commodity with the special capacity for symbolizing the value inherent in all other commodities. Smith's account thus stands at the heart of the so-called commodity theory of money, which I have discussed elsewhere.[3] The most immediate problem with this account is that, as formulated by Graeber, 'there's no evidence that it ever happened, and an enormous amount of evidence suggesting that it did not'.[4] The account of money as emerging out of a barter economy is anthropologically and historically wrong.[5] Primitive communities were rarely if ever organized as barter economies. Instead, they were organized in ways where members of the community would share resources and labour and offer these as gifts to the community and to each other.[6]

There are several philosophical implications of this conception of the origin and function of money, which not only pertain to classical but also to much of neo-classical economics. We shall be exploring some of these implications in the current as well as following sections. The first thing to note is that Smith's account of money tends to veil the function of money in the very constitution of the economic subject. The butcher, the brewer and the baker are already economic subjects well versed in the 'truck, barter, and exchange' of a market economy. The introduction of money into the economy does nothing but facilitate more efficiently the form of exchange that is already taking place. Eco-analysis obviously disagrees on this point. As we have discussed previously, Smith is correct in noting that the division of labour makes the labourer dependent on the market for the provision of produce to satisfy his needs and wants. But the role of money is not only, as suggested by Smith, to facilitate a more efficient exchange of commodities and services in the market. Money itself plays a curious role in the constitution of the economic subject. The introduction of money signifies the economic castration of the subject. In order to see what that means, we shall look into Žižek's general theory of the relation between desire and objects:

> [T]he subject... and the object-cause of its desire... are strictly correlative. There is a subject only in so far as there is some material stain/leftover that *resists* subjectivation, a surplus in which, precisely, the subject *cannot* recognize itself. In other words, the paradox of the subject is that it exists only through its own radical impossibility, through a 'bone in the throat' that forever prevents it (the subject) from achieving its full ontological identity.[7]

Commodity objects in the market function as the object-cause of the desire of the consumer. But money itself takes on a strangely ambivalent character. On the one hand, money is the most general expression of desire. Since money may be exchanged for every kind of object in the market, money carries the capacity to satisfy any kind of desire. Money is the object that levels out all qualitative differences between individual consumers. On the other hand, money is the most useless of all the objects in the market. Money itself does not have any use-value. In this sense, money itself does not correspond to any particular desire in the consumer. We see here how money fits into Žižek's account of the relation between subject, desire and object. Money is the 'leftover that *resists* subjectivation, a surplus in which, precisely, the subject *cannot* recognize itself'. While the consuming subject may identify with different objects in the market – 'I need a cup of coffee right now'; 'this dress goes well with the colour of my eyes'; 'this resort is perfect for our honeymoon' – as these objects are imagined to enable the subject to 'achieve its full ontological identity', it makes much less sense to identify with money as such. Since money has no immediate use-value, it is not correlated with any desire in the subject. Yet at the same time money is the thing that we all desire. While the desire for different particular commodity objects is what defines my identity, the desire for money stands for 'that in me that is more than myself'. In other words, money is what Lacan refers to as *objet petit a*. Žižek provides the following account:

> Materialism means that the reality I see is never 'whole' – not because a large part of it eludes me, but because it contains a stain, a blind spot, which signals my inclusion in it. Nowhere is this structure clearer than in the case of Lacan's *objet petit a*, the object-cause of desire. The same object can all of a sudden be 'transubstantiated' into the object of my desire: what is to you just an ordinary object, is for me the focus of my libidinal investment, and this shift is caused by some unfathomable x, a *je ne sais quoi* in the object which cannot ever be pinned down to any of its particular properties.... *L'objet petit a* can thus be defined as a pure parallax object.... The paradox is here a very precise one: it is at the very point at which a pure difference emerges – a

difference which is no longer a difference between two positively existing objects, but a minimal difference which divides one and the same object from itself – that this difference 'as such' immediately coincides with an unfathomable object: in contrast to a mere difference between objects, the pure difference is itself an object.[8]

This very dense passage provides several theoretical conceptualizations to explore the role of money in the constitution of the economic subject. Let us invoke once again the distinction between the law of equivalent exchange and the law of impossible exchange. If we think of the economy merely in terms of the law of equivalent exchange, the desire for money is ultimately backed by the desire for the commodities and services that may be bought for money. The value of money ultimately resides in the exchange-value of money, that is, money as a medium of exchange for commodities of equal value. However, the desire for money cannot be reduced to the underlying desire for the commodities. If we understand the economy merely in terms of the law of equivalent exchange, we fail to see the particular role of money and we fail to see 'the whole' of the economy. Once we shift our perspective to the law of impossible exchange, we can see that money is in fact 'a stain, a blind spot, which signals [the subject's] inclusion in' the economy. In other words, the desire for money is ultimately the desire to be included in the economy as a whole.

While the exchange of commodities for money is based on the law of equivalent exchange, the process by which desire is subjected to the sphere of economic exchange is an instance of impossible exchange. Paraphrasing Žižek's definition of symbolic castration: the incorporation of desire into the symbolic order of economy and equal exchange is the exchange of 'something which the subject never possessed in the first place' for 'a purely potential, nonexistent X, with respect to which the actually accessible experiences appear all of a sudden as lacking, not wholly satisfying'. The desire of the subject is exchanged for money in the sense that the subject begins to express its desires in terms of objects that may be acquired in exchange for money. In this kind of exchange, desire takes the form of demand and the satisfaction of this demand follows the law of equivalent exchange as the price of satisfaction is determined according to the interaction between supply and demand. However, there is also another dimension to the exchange of desire and money since the expression of desire in terms of purchasable commodity objects also creates a fundamental lack that cannot be fulfilled by any positive object. In this context, money itself appears as the *objet petit a*, the very cause of desire itself. The desire caused by money itself cannot be exchanged for money. This is again where the economy comes up against the barrier of impossible exchange. The desire for money itself does not correspond to any pre-existing need outside of the economy and therefore it cannot be satisfied through any form of exchange.

Part II touched upon the way that money 'divides one and the same object from itself' as it institutes the difference between price and value in the commodity. Price is the numeric symbol that signifies the amount of money at which the commodity is exchanged in the market. In other words, money is the object that creates the 'parallax gap' between price and value in the commodity. What we have now seen in relation to symbolic castration is that money also functions to 'divide one and the same *subject* from itself'. As the subject is incorporated into the symbolic order of economy, it emerges as the split subject. In Lacan's system of notations, that split subject is denoted by the symbol $. The similarity with the dollar sign, which stands as the eminent symbol of money, is obvious. The concept of the split subject refers to the traumatic constitution of the subject. In Žižek, subjectivity does not precede the process of symbolization. Subjectivity only emerges in the very process of symbolization and then always already as an impossible compromise formation.

In the context of ecology, the split subject emerges as it is 'included out' of the domain of nature governed by the inherent balance of nature. In the context of economy, the subject is perhaps rather 'excluded in' as the economy points to the natural needs of the subject, the satisfaction of which is the ultimate purpose and legitimacy of the economic system. The subject comes to function as the *point de capiton*, where the symbolic order of economy is interstitched into the fabric of the real. This was exemplified by Brundtland's reference to the 'essential needs of the world's poor'. But, at the same time, the institution of the economy also has the effect of subsuming all the different needs of economic subjects into a common desire for money that has no correlate in the alleged natural constitution of the subject. On the one hand, the economy appears as nothing but a very efficient system for the production and distribution of goods that serve to satisfy human needs. On the other hand, the economy produces the desire for money as perhaps the most human of all human needs. The desire for money transcends the desire for any particular object. '[D]esire's *raison d'être*...is not to realize its goal, to find full satisfaction, but to reproduce itself as desire.'[9] This element of pure reproduction of desire is incarnated in the desire for money.

The subject of economy fits the figure of the split subject. This is for instance illustrated by the neo-classical production function, where the subject simultaneously figures on both sides of the equation, which is divided by the '='. On the one hand, the subject is labour that is producing commodities exchanged for money in the market. In the production function, this is symbolized as L on the right-hand side of the equation. On the other hand, the subject is a consumer exchanging money for objects of consumption in the market. Consumption is symbolized as C on the left-hand side of the equation. Things that flow directly from production to consumption are not actually registered as economic output. They transgress the division between production and consumption, thus flying under the radar of economics. Classic examples include home-grown carrots, swimming in the sea, domestic child care and sexual intercourse. In all of these cases, valuable goods or services are produced and consumed and demands for nutrition, entertainment, care or sex are met. However, satisfaction of the demands is not mediated through monetary exchange in the market but rather in a more immediate fashion, eluding the domain of the monetary economy. The production and consumption of these goods and services does not contribute to the growth of the economy. Instead, the distinction between producer and consumer is blurred in a way that undercuts the split subject of the economy. *Homo economicus* is a subject unable to satisfy his own desires outside of the domain of monetary exchange. The division of labour inherent in the constitution of the economy is not only the division of labour from itself but also the division of labour from consumption. The subject of economy is split between his or her identity as producer and his or her identity as consumer.

With regard to the concept of green growth, the point here is that, by retaining the aspiration of growth, political visions which incorporate this concept remain trapped within a very narrow understanding of economy because they fail to question the constitution of the subject as a consumer. Policies of green growth may involve a call for the consumer to eat biodynamic bananas and drive electric cars, but the existing call for ever more consumption is left intact. We have seen how Žižek explains symbolic castration by way of the prohibition of incest. In the context of the economy, we find a similar prohibition at play. This is the prohibition against autarky: you must not enjoy the fruits of your own labour! The concept of green growth remains well within the boundaries of this prohibition. Autarky means self-sufficiency. An autark is someone who remains independent of the larger economy as he is able to produce and provide for himself everything he needs. This means that the autark does not enter into relations of exchange in order to buy and sell labour, goods and services. The autark stands as the diametrical opposite to *homo economicus*. Autarky constitutes a point of indistinction, where the difference between production and consumption breaks down. When a labourer is included in the symbolic order of the economy by receiving money in exchange for his labour, he is simultaneously provided with buying power that may be used to purchase commodities and services in the market. He thus figures twice in the economy function: as labourer and as consumer. It is crucial to note how the relation between the two positions of the labourer/consumer is mediated by money. The labourer needs to work, not in order to produce objects to satisfy his needs and desires but in order to earn money to buy objects to satisfy his needs and desires. The inclusion of the labourer in the symbolic order of the economy is not merely a matter of division of labour, whereby production processes are rationalized by being split into partial components performed by different labourers, each adding their little contribution to a larger whole. It is also a matter of a division between production and consumption. The labourer works not in order to produce the things he needs for the maintenance of himself and his household. He works to earn money that may be used to buy the things he needs for the maintenance of himself and his household.

For the autark, the relation between production and consumption is not mediated by money, and so s/he does not figure in the economy as labourer or as consumer. The autark simply eludes the kind of symbolization performed as we perceive the economy through such models as the neo-classical production function or even just the registering of economic activity through GDP. Of course, the prohibition of autarky does not mean that we cannot grow our own vegetables or have sex with our own spouses. It works through a logic of exclusion. It is not that autarkic production and consumption, eluding monetary exchange in the market, is forbidden. It is just not considered to be part of the economy. This means that we have certain forms of production and consumption that do in fact function to satisfy certain needs and demands, but they do not figure as part of the domain of the economy. The concept of green growth fails to register and recognize the existence, function and potential of such forms of production and consumption. And, moreover, it comes to stand in the way of a critical examination of the constitution of the consumer.

An interesting case to illuminate this issue is provided by campaigns to reduce food waste. The poster shown in Figure 7.1 is from the 'Love Food Hate Waste' campaign, which was initially launched in the United Kingdom and subsequently spread to other countries. Such campaigns are interesting because they operate on the margins

The Desire to Grow 171

Figure 7.1 'Love Food Hate Waste' campaign poster

of the concept of green growth. On the one hand, the call for a reduction of food waste may seem like nothing but a sensible and modest piece of consumer advice: 'Reduce the amount of food you waste by shopping to a list and buying only what you need.' This is hardly a call for a world revolution against capitalism. On the other hand, campaigns to reduce food waste are remarkable as they not only address people in their capacity as consumers but simultaneously as producers. Advice on how to reduce waste extends to providing recipes for meals made of leftovers, guidelines for meal portions, tools for meal planning and so on. Essentially, campaigns against food waste aim to reinvigorate good old-fashioned virtues of housekeeping. In some accounts of the etymology of the word economy, *oikos* is translated into household and *nomos* into management, whereby 'economy' is found to be derived from 'household management'. If we stick with this interpretation, food-waste campaigns are all about promoting good principles of economy. But this also means that they have the potential to expand our understanding of the domain of the economy beyond the narrow boundaries of the market economy where goods and services are exchanged for money.

Campaigns against food waste thus have the potential to point beyond the paradigm of green growth. If the domestic consumer/producer actually succeeds in reducing food waste by adopting more 'economic' patterns of shopping as well as food preparation and conservation, it could reduce the turnover of the food retail sector and thus also of the volume of the food production industry. The campaigns would thus lead to more 'green' but not more 'growth'. In this light, it is remarkable how some campaigns have managed to engage major food retailers as sponsors and collaborators in the effort to reduce food waste. Whether such engagement is an expression of altruistic commitment to the cause or simply a marketing strategy to gain market shares from competitors should be judged by the kind of changes retailers are willing to implement. It is one thing to sponsor a campaign and to have one's brand name on the campaign website. It is quite another thing to stop quantity discounts, offer 'imperfect' fruit and vegetables on the shelves, sell products nearing 'best before' dates at a discount, cut down on customer marketing or implement other initiatives that might come at a cost to the overall turnover of the retailer.

And if we pursue the problem of food waste even a few steps further than most of the official campaigns, we might begin to question the very organization of our contemporary food industry. Perhaps the problem is not merely a matter of shopping behaviour, meal planning, labelling and discounts. The very organization of the food industry on the basis of a global division of labour obviously leads to an increasing divide between the consumer subject and the means of production of the commodities necessary for his or her subsistence. Perhaps this is the more fundamental explanation for our careless and disrespectful interaction with food, which leads to its massive waste. We shall come back to the issue of food but first we need to look further into the concept of labour.

Green Growth and Green Jobs

If we look into political calls for green growth such as the ones from the Danish government, US President Obama and the OECD quoted at the beginning of this chapter, they are typically coupled with the aim of 'creating new jobs'. The means to achieve green growth is through innovation and new knowledge, which in turn also provides a competitive advantage in the global market. In this sense, the vision of green growth is perfectly aligned with the neo-classical fantasy of perpetual growth through 'knowledge or technology'. The so-called 'cleantech' industry is already a booming sector and is regarded as having an even greater potential for growth in the future. Here is how US President Obama expresses some of these visions:

> Over the past four years, we've doubled the electricity that we generate from zero-carbon wind and solar power. [Applause.] And that means jobs – jobs manufacturing the wind turbines that now generate enough electricity to power nearly 15 million homes; jobs installing the solar panels that now generate more than four times the power at less cost than just a few years ago. /.../ [C]ountries like China and Germany are going all in in the race for clean energy. I believe Americans build things better than anybody else. I want America to win that race, but we can't win it if we're not in it. [Applause][10]

It is interesting to note how solving the global issue of climate change is intertwined with purely national interests. America must save the planet while at the same time beating its competitors in the global market. This creates profits for US companies and jobs for US workers. The conflation of the global and the national is in no ways unique to the United States. We find similar formulations in the proclamations of other countries that have a policy on green growth. Green growth to save the planet goes hand-in-hand with the creation of 'green jobs' in the national economy. Here is how the vision is spelled out by the EU:

> Europe 2020, the EU's new economic strategy, stresses the need for smart, sustainable and inclusive growth. That means building a competitive, low-carbon, resource-efficient economy and safeguarding the environment. The main goal under the Europe 2020 strategy is to support businesses and to enable them to improve their competitiveness globally whilst helping them make the shift towards a green economy. /.../ More than 20 million European jobs are already linked to the environment in some way – and as the EU gears up for a greener future, we could see the creation of millions more green jobs.[11]

Now the eco-analytical question is not whether the creation of green jobs through green growth is in and of itself good or bad or even whether green jobs are better than 'brown' jobs. It is arguably

Figure 7.2 Social Democrats' campaign poster

better to have people work to build windmills than to have them extract oil from tar sands. But maybe this is a false choice. The concept of green growth tends to take for granted that growth is needed in order to 'create new jobs'. This is the standard argument for perpetual growth in mainstream left-wing political parties and even in most parties on the right. The economy must grow in order to solve the problem of unemployment. Figure 7.2 shows a 2014 European Parliament election campaign slogan for the Danish Social Democrats. The text on the helmet reads: 'More Green Jobs Now'.

This claim might be scrutinized on the level of economics itself. Is it really true that economic growth leads to less unemployment? Or could we argue that economies today grow in ways that do not necessarily translate into more employment? But even with these kinds of questions, we are not on the course of eco-analysis. On the contrary, we need to backtrack from the concept of labour itself in order to understand the relation between jobs, growth and economy. The question is not: How can we create more jobs through green growth? But rather: What is the difference between labour and jobs, and why do we need more jobs in the first place? The following analysis looks into yet another dimension of the division of labour. This is the division of labour from itself, which is at the heart of the division between labour and jobs.

Let's start at the beginning. If we return to the context of the Bible and the expulsion of Adam and Eve from the Garden of Eden, we can identify a division of labour preceding the kind of division of labour of which Adam Smith speaks. This is the division of labour from itself, which is instituted by the Lord himself. As God discovers how Eve has managed to tempt Adam into eating the forbidden fruits from the Tree of Knowledge, He is furious and so casts an eternal curse on both Adam and Eve:

> He told the woman, 'I'll greatly increase the pain of your *labor* during childbirth. It will be painful for you to bear children',...He told the man, 'Because you have listened to what your wife said, and have eaten from the tree about which I commanded you "You must not eat from it," cursed is the ground because of you. You'll eat from it through pain-filled *labor* for the rest of your life. It will produce thorns and thistles for you, and you'll eat the plants from the meadows. You will eat food by the sweat of your brow until you're buried in the ground, because you were taken from it.' (Genesis; my italics)

What is interesting to note in our current context is the way that God condemns Eve and Adam into performing two different but both painful forms of labour. There is the labour of childbirth, the production of children from the womb of the mother. And there is the labour of cultivating the land into producing food, which is the kind of work that ultimately results in the production of commodities and services for the market. Even though labour today still maintains this double meaning of giving birth and working, its two forms do not figure on equal terms in our current conception of economy and certainly not in the neo-classical conception of the economy. We can understand the divine division of labour into these two forms in terms of our distinction between the two laws of exchange.

Even though pregnancy is perhaps one of the most sophisticated forms of production known to mankind, maternal labour is not registered as part of the productive capacity of the economy. In fact, we do not typically use the word 'production' with regards to the making of a baby. Instead, we speak of 'giving birth'. The birth of a child is conceived as a gift rather than a form of production for which something is provided in return. The birth of a child is not subsumed under the law of equivalent exchange. To whom would something be given in return, if we were to conceive the birth of a child as one side of an equivalent exchange? The mother, the father, God or perhaps the child itself? Who would pay for the birth of the child? And what could possibly be given that would equal the life of a human being? Of course, there are great historical and cultural differences in the ways that we conceive of life, birth and children, and there are many different ways in which the birth of a child is inscribed into religious and economic systems of exchange. Rather than digressing into an anthropological or even theological study of the economics of kinship, the point here is merely to assert that the economy conceived by modern economics in general and neo-classical economics in particular does not register the birth of a child as adding to productive output. If anything, the birth of a child is registered as a decrease in GDP per capita in so far as there is now one more mouth to feed in the economy. It is only when the child becomes part of the labour force that it is registered as anything but a liability. This is of course when the child is capable of performing the kind of labour that results in the production of commodities and services for the market. This is the kind of labour that is subsumed under the law of equivalent exchange. The division of labour from itself is inscribed into the neo-classical conception of the economy as it constitutes the very boundary of the domain of the economy. Maternal labour and the resulting birth of a child mark the point where productive labour comes up against the law of impossible exchange and falls outside the scope of the economy.

This division of labour from itself is not only relevant to the specific case of giving birth to children. It is merely the most emblematic example of a great number of labour activities that are productive but not registered as part of economic output in society. We have already touched upon such examples as domestic child care and sexual intercourse, which are of course both closely related to the making of children. The list of examples is endless and includes everyday activities such as: the processing and preparation of food to be consumed among family and friends; the repair of things that would have otherwise been discarded; the teaching of a skill to a child; the entertainment of other people through the telling of a story; the giving of an advice to a friend, relative or even a stranger; or the care of someone who is ill. What characterizes these different activities is that they are forms of labour that, even though we would probably hesitate to call them labour, are provided without monetary compensation. They are acts of labour which are not jobs. They are not governed by the law of equivalent exchange but rather by the (il)logic of a gift economy.

In part II, we saw how the neo-classical circular flow diagram of the economy veils the interaction between the economy as a whole and the real of the eco. The extraction of natural resources, as well as the discarding of waste back into the natural environment, does not show up in the production function. A similar mechanism is at play with regard to labour. Only paid labour, which is subject to the law of equivalent exchange by being compensated in money, is registered as input into the economy and thus also as contributing to the volume of productive output. This conversion of paid labour into marketable goods and services respects the division of the subject into producer and consumer. Labour, however, that is not compensated in money but rather offered as a gift violates the law of equivalent exchange and also does not respect the division of the subject into producer and consumer. This kind of labour does not show up in the neo-classical account of the economy. With Žižek, we can also think of the difference between labour and job as the difference between the real and the symbolic. A job is a labour activity that is symbolized through the payment of a salary and integrated into the symbolic order of an organization such as a company. Labour that is not symbolized as a job is barely recognized as labour. It is difficult to register the value of a labour activity that is not exchanged for money, thus rendering a particular price.

When a mother breastfeeds a baby or a father teaches his son how to ride a bicycle, it makes no sense to ask who the producer is and who the consumer is, who is giving and who is receiving. Even though it might seem that the mother and the father are the ones providing something for their children, the relations are reciprocal as the very act of giving creates a mutual bond that is immediately valuable to both parties. Giving a gift creates a value without a price. This extends beyond relations between parents and children. In a community, where people readily help each other and extend favours, social bonds are strengthened in a way that cannot be reduced to individuals owing each other certain services. Opposed to the exchange of labour for money, which we find in the economy, the giving of gifts belongs in the order of the real. This also means that the kind of labour offered as a gift is a kind of real labour that operates according to the law of impossible exchange. The notion of 'exchanging gifts' is an oxymoron. A pure gift is offered without any expectation of getting something in return other than the simple joy of giving the gift.

Compare the following two scenarios: (1) As Bent is clearing away the snow in front of his own house in the morning, he sees that the pavement in front of the house of his neighbour, Rita, is also full of snow. He knows that Rita has a bad hip and also does not like hard physical labour, so he also clears the snow in front of her house. In the evening, Rita is making a casserole. Since she has made too much food for herself to eat anyway, she invites Bent to come and join her for dinner. (2) Bent has a small business that provides gardening and related services to private customers. Rita has a bad hip, so when the winter comes she pays Bent's company 200 Danish kroner to clear the snow in front of her house. Rita makes the money running a bistro. Since Bent has been clearing snow all day, he is too tired to cook so he decides to go to the bistro for dinner. He orders the meal of the day, which is a casserole and a drink, priced at 200 kroner. In the second scenario, the two transactions cancel each other out. Each party pays money for the service of the other party and at the end of the day they are even. From a purely economic perspective, the same appears to be the case in the first scenario. But although Bent and Rita are formally even, it seems reasonable to assume that the two acts of gift giving have functioned to form some kind of affective relation between them.

In part II, I explored how the neo-classical conception of economics veil the natural origin of matter that comes to constitute capital in the economic system of production. Under the Lacanian paraphrase 'nomy kills the eco', this was analysed in terms of the way that the economic process of symbolization effaces the real of the eco. This point also applies to the transposition of real labour offered as a gift to economic labour in the form of a job that is compensated according to the law of equivalent exchange. From a purely economic perspective, the transposition of gift giving into market-based exchanges mediated by money may appear to constitute nothing but a more efficient way of organizing the interactions between members of a community: the pricing of a product of labour and the subsequent exchange of this product for an amount of money with the capacity of buying the product of someone's labour is nothing but a simple representation of the intrinsic value of the product. Such understanding of money as nothing but a means to facilitate the exchange of one commodity for another is sometimes referred to as the 'veil of money'. This concept, derived from classical economics, means that behind the 'veil of money' the economy is ultimately a barter economy. John Stuart Mill, for instance, remarks:

> There cannot, in short, be intrinsically a more insignificant thing, in the economy of society, than money; except in the character of a contrivance for sparing time and labour. /.../ The introduction of money does not interfere with the operation of any of the Laws of Value laid down in the preceding chapters. /.../ Things which by barter would exchange for one another, will, if sold for money, sell for an equal amount of it, and so will exchange for one another still, though the process of exchanging them will consist of two operations instead of only one. The relations of commodities to one another remain unaltered by money.[12]

Eco-analysis obviously disagrees with this concept of money. Not only are the 'relations of commodities' indeed altered by money because certain kinds of capital, such as natural capital, are systematically excluded from the domain of economics but also the relations of subjects to one another are altered by the introduction of money. The introduction of money converts relations of gift giving or sharing into relations subsumed by the law of equivalent exchange. The concept of the 'veil of money' plays the same ideological role as Smith's account of the origins of money that we have already explored. In order to unfold the role of money in the division of labour from itself, let's review a contemporary version of Smith's account. The following passage comes from a textbook in economics where it figures in a chapter entitled 'What is Money?':

> The use of money as a medium of exchange promotes economic efficiency by minimizing the time spent in exchanging goods and services. To see why, let's look at a barter economy, one without money, in which goods and services are exchanged directly for other goods and services.
>
> Take the case of Ellen the Economics Professor, who can do just one thing well: give brilliant economics lectures. In a barter economy, if Ellen wants to eat, she must find a farmer who not only produces the food she likes but also wants to learn economics. As you might expect, this search will be difficult and time-consuming, and Ellen might spend more time looking for such an economics-hungry farmer than she will teaching. It is even possible that she will have to quit lecturing and go into farming herself. Even so, she may still starve to death.
>
> The time spent trying to exchange goods or services is called a transaction cost. In a barter economy, transaction costs are high because people have to satisfy a 'double coincidence of wants' – they have to find someone who has a good or service they want and who also wants the good or service they have to offer.
>
> Let's see what happens if we introduce money into Ellen the Economics Professor's world. Ellen can teach anyone who is willing to pay money to hear her lecture. She can then go to any farmer (or his representative at the supermarket) and buy the food she needs with the money she has been paid. The problem of the double coincidence of wants is avoided, and Ellen saves a lot of time, which she may spend doing what she does best: teaching.[13]

First of all, we see how the parable of Ellen the Economics Professor reproduces the same fantasy of a pre-monetary barter economy that we found in Smith's story of the butcher, the baker and the brewer. Second, the parable provides an eminent example of the way that ideology, according to Žižek, functions not only by providing 'wrong answers' but even 'wrong questions'. Along these lines, we can see how the way that the parable perceives the problem is itself part of the problem. The problem as posed in the passage is this: 'Take the case of Ellen the Economics Professor, who can do just one thing well: give brilliant economics lectures. In a barter economy, if Ellen wants to eat, she must find a farmer who not only produces the food she likes but also wants to learn economics.' If, for a moment, we suspend the ideology of economic abstraction and read the problem using common sense, we can see a much simpler, much more realistic, but then of course also much more politically incorrect, solution: 'Take the case of Ellen, who can do just one thing well. In a barter economy, if Ellen wants to eat, she must find a farmer she likes.' This is how things would work out in an actual economy without money. In the real world, there is no such thing as 'economics-hungry farmers'. However, there are plenty of farmers who like women and sex and who want to make a family. Ellen would thus enter into a relation with a farmer. They would make love, she would give birth to his children and he would of course share with her the fruits of his labour. Their relation would not be one of barter, where sex and babies were in some way exchanged for food and shelter according to the law of equivalent exchange. Instead, their relation would probably be inscribed into some form of gift and sharing economy.

The account of money found in classical as well as neo-classical economics functions to project elements of the current state of affairs back into the past as to present these as trans-historical fundamentals. Once money has emerged and has come to function as the main structuring principle of economic exchange, it retroactively creates a fantasy of pre-monetary exchange. The account of money as emerging out of a primitive barter economy projects the fantasy that, even before the emergence of money, economic exchange was structured by the law of equivalent exchange. In the parable of Ellen the Economics Professor, we see how this projection works with respect to the division of labour into different forms of job such as farmer or economics professor. This form of division of labour is instituted as preceding the emergence of a money economy. What is obscured through this projection is of course the way in which such a division of labour only emerges with the introduction of money. Furthermore, the division of labour from itself is wholly effaced in the classical and neo-classical account of the economy. The forms of labour that are not organized as jobs and not performed with the purpose of producing commodities and services for the market are not registered as part of the economy because they operate according to the logic of the gift.

Gift giving is the real of economy in the sense that it constitutes an implosion of the conventional rules of economic exchange. This is why 'gift economy' is perhaps an oxymoron. The gift belongs in the order of the eco. While theft is a simple violation of the law of equivalent exchange, the gift is a displacement into the order of the law of impossible exchange. The gift is that which 'resists symbolization' through the law of equivalent exchange. The fantasy of barter as the original pre-monetary exchange is a projection to protect the economy from the traumatic event of the gift. The symbolization of exchange through money is not just a simple update of an already existing form of exchange but rather a violent interruption of non-monetary forms of interaction. We may add another dimension to the idea that 'nomy kills the eco' by saying that *money kills the gift*.

This brings us back to the ambivalent nature of economic growth. An increase in GDP represents an increase in the production of commodities and services exchanged for money in the market. We cannot know, however, whether this economic growth means that demands which were previously unmet are now being served by the market economy, or whether it means that demands which were previously being met outside of the market are now being served by the market economy. A similar ambivalence applies to the issue of labour. When new jobs are created through economic growth, it means that more people and more labour are being integrated into the market economy. We cannot know, however, whether this is an integration of labour that was previously idle, or whether it is an integration of labour that was previously performing productive functions outside of the market economy. Job creation through economic growth is the incorporation of labour under the law of equivalent exchange. Since labour that is performed according to the law of impossible exchange is not registered as part of the economy, we cannot tell if the growth of the former kind of labour takes place at the cost of the latter kind. We cannot tell if *jobs kill labour*.

Adam Smith and Eve

At the heart of the division of labour between Adam and Eve is of course the controversial issue of gender. In order to explore this issue, let's look at a curious parallel between the environmental movement and the women's movement. We have already discussed how sustainability signifies a *new spirit of capitalism*.[14] The emergence of this spirit of green capitalism is a perfect illustration of the dialectics between capitalism and the critique of capitalism as outlined by Boltanski and Chiapello. The evolution of the environmental movement in the 1960s and the 1970s sparked a growing awareness of the ecological consequences of the capitalist mode of production. But rather than leading to the overthrow of capitalism as such, this critique has been recuperated by capitalism and thus used to fuel the wave of green capitalism that has now become part of the mainstream.[15]

In similar fashion, the so-called second-wave feminism of the 1960s and the 1970s also launched a severe critique of capitalism, pointing to inherent structures of gender inequality in the capitalist mode of production. A central argument in the women's movement at the time was that capitalism was centred on wage work, thus failing to recognize and valorize the different forms of unpaid care work within the household. Since this form of work was typically carried out by women, it allegedly left them disadvantaged in relation to their wage-earning men. But, again, rather than leading to any subversion of the capitalist mode of production, the second-wave feminist critique has also been recuperated by capitalism, where it dovetails nicely with a spirit of neo-liberal capitalism that has gained momentum since the 1980s. Contemporary feminism is less concerned with the decentring of wage work and the valorization of unwaged labour and more concerned with issues of equality that are measured by the capitalist criteria of the market. Here is how Nancy Fraser sums up the merging of feminism and neo-liberalism:

> Disturbing as it may sound, I am suggesting that second-wave feminism has unwittingly provided a key ingredient of the new spirit of neoliberalism. Our critique of the family wage now supplies a good part of the romance that invests flexible capitalism with a higher meaning and a moral point. Endowing their daily struggles with an ethical meaning, the feminist romance attracts women at both ends of the social spectrum: at one end, the female cadres of the professional middle classes, determined to crack the glass ceiling; at the other end, the female temps, part-timers, low-wage service employees, domestics, sex workers, migrants, EPZ workers and microcredit borrowers, seeking not only income and material security, but also dignity, selfbetterment and liberation from traditional authority. At both ends, the dream of women's emancipation is harnessed to the engine of capitalist accumulation. Thus, second-wave feminism's critique of the family wage has enjoyed a perverse afterlife. Once the centrepiece of a radical analysis of capitalism's androcentrism, it serves today to intensify capitalism's valorization of waged labour.[16]

The Nordic countries are sometimes regarded as the vanguard of gender equality and women's liberation. It is also among Nordic countries such as Iceland, Sweden and Denmark that we find the highest occupational rate of women in the world.[17] This indicates how women's liberation has come to be synonymous with inclusion and self-realization in the market economy. Such conception of liberation is perfectly aligned with the spirit of neo-liberal capitalism that promotes the logics and measures of the market to more and more domains of life. This alignment is not only ideological. It has very concrete econometric effects. The incorporation of an increasing number of women in the labour market contributes to economic growth. The labour performed by a mother taking care of her own children within the household does not count as a contribution to GDP since no marketable goods or services are produced. But when this woman is employed in a paid job, for instance in a kindergarten taking care of other people's children, it is priced and registered as part of the nation's productive output. And if her own children are at the same time being taken care of by another woman in another kindergarten, the pay-off is double. Of course, the organization of care work within the welfare state or the commercial enterprise rather than the individual household also offers some economies of scale which free labour power to provide growth in other sectors of the economy. Levelling out the kind of division of labour that was instituted between Adam and Eve facilitates the efficiency gained through the kind of division of labour that Adam Smith is talking about.

In our introductory discussion of the etymological roots of the concept of 'economy', we briefly touched upon the affinity between *oikos* and 'household'. We also saw how *nomos* could be interpreted as 'management'. If we were to go along with these, rather than the much broader interpretations of *oikos* as 'habitat' and *nomos* as 'naming', we might conceive of the concept of economy as originally meaning 'household management' or simply 'housekeeping'. This is an etymology that is sometimes invoked in the field of ecological economics.[18] This etymology reveals yet another inherent paradox in the pursuit of perpetual economic growth. As the economy grows in conventional terms as measured by GDP, there seems to be an inverse tendency for the 'economy' in the narrow interpretation of housekeeping to de-grow. To the extent that conventional economic growth has presupposed the inclusion of more and more women in the labour market, this leaves less time and less immediate need for the kind of labour required to maintain good housekeeping. Rather than preparing meals from basic ingredients, ready-made meals are purchased outside of the home. And also the caring for children and elderly is being outsourced to public or private enterprises that are part of the market economy.

The inverse relation between the two economies also plays out in the dimension of knowledge and education. While the level of formal education has indeed been steadily rising in most western countries over the past several decades, the knowledge and skills learned in universities and most other educational institutions are typically targeted at what is demanded by employers in the labour market. Especially with the neo-liberal reforms that have been sweeping through the educational systems in recent years, the purpose of education is increasingly to become the object of the desire of the labour market. Wendell Berry argues that American universities are structured simply as one great 'major in Upward Mobility': 'The Upward Mobility major has put our schools far too much at the service of what we have been calling overconfidently our "economy." Education has increasingly been reduced to job training, preparing young people not for responsible adulthood and citizenship but for expert servitude to the corporations.'[19]

In similar fashion, as GDP measures only those goods and services that are produced for the market, so there is also a tendency in the formal education system to train and value only those skills that are required in various job functions in the labour market. In our current 'knowledge society', it is close to blasphemy to argue against knowledge and education as such. Our commitment to the universal and unconditional value of formal education is illustrated by the fact that one of the key justifications for the war in Afghanistan ended up being that it was necessary in order to ensure that Afghan girls could go to school. Education is apparently so important that we are willing to go to war for it. Without endorsing Taliban educational policies, we might pose the critical question if the increased focus on academic and professional skills in our contemporary knowledge society simultaneously results in a de-skilling of other competencies. These might be not competencies demanded in the labour market but nevertheless required in the 'housekeeping economy'.

There is a certain irony in the organization of university education on large campus facilities found in many places in the United States, as well as in several other countries. These are the places where young people are going to learn essential skills required to navigate the world for the rest of their lives. At the same time, such campus facilities are typically secluded from any 'ordinary' life environment. Students live far away from their families and native communities. They do not have to worry about cooking, cleaning, repairing, constructing or even caring about anyone other than themselves. And they do not have to relate to very many people who are different from themselves in terms of class, age or way of life. They are, in other words, 'free' to concentrate solely on the achievement of academic skills, such as the ability to deconstruct political texts, discuss realism versus idealism or make regression analyses of consumer preferences. Rather than giving young people a set of general life skills, university education tends to constitute yet another dimension of the economic castration that we have been exploring. After five years of college education in the gated community of a university campus, graduates will have become even more dependent on the kind of jobs offered by the corporate world or the state bureaucracy since these are the only institutions capable of valorizing the highly specialized and abstract competences on which the productivity and self-worth of these graduates have come to rely.

Debates about the educational system typically centre on the question of whether universities and other learning institutions provide students with the right kind of skills demanded by the labour market. In times of recession, these debates are further fuelled by students who are understandably worried about whether their education will provide them with future employment: does university education make graduates the object of desire of the big Other of the labour market? Such debates, however, typically stand in the way of more fundamental reflections on the nature, functions and effects of academic education. It is for instance interesting to speculate about what five years of university education does for a person's life skills in relation to the building and maintenance of a household. Do women become better or worse at being wives, mothers, lovers, housekeepers, neighbours and citizens? Do men become better or worse at being husbands, fathers, lovers, housekeepers, friends and citizens? Or is it perhaps impossible to say anything general about this issue?

These questions add another dimension to the discussion of food waste in the previous section. Estimates say that about half of global food production ends up as waste rather than human food.[20] Now, of course, it would be grossly reductionist to blame this shameless handling of food solely on university education. However, it does seem fair to suggest that the figure indicates some kind of deterioration in basic housekeeping skills. Such skills include the 'economic' preparation, planning and conservation of food in the household. But they also include the 'economic' use and preservation of natural resources in the primary production of produce. Housekeeping skills are rarely required in the performance of ordinary salaried jobs and are thus not demanded by the labour market. Hence they are rarely part of the kind of knowledge that is being valued and fostered in the 'knowledge society'.

The issues of gender and education in relation to economic growth are of course not identical. Nevertheless, they are related since both involve inconvenient truths that come up as we delve further into the foundations of our contemporary growth imperative. Breaking with this imperative is not just about recycling waste, buying organic milk or skipping holidays to Thailand. Women's liberation and the increase in the average level of formal education are generally conceived as unconditional goods, and they are taken as signs of progression in western civilization. But what if these ideas have today become structured in a way that makes them inherent parts of the ideology that keeps us tied to the compulsion of perpetual economic growth? Berry provides the following observation that may serve to illuminate the question:

> Recently I heard, on an early-morning radio program, a university economist explaining the benefits of off-farm work for farm women: that these women are increasingly employed off the farm, she said, has made them 'full partners' in the farm's economy. Never mind that this is a symptom of economic desperation and great unhappiness on the farm. And never mind the value, which was more than economic, of these women's precious contribution *on* the farm to the farm family's life and economy – in what was, many of them would have said, a full partnership. *Now* they are 'earning 45 percent of total family income'; *now* they are playing 'a major role'. The 45 percent and the 'major role' are allowed to defray all other costs. That the farm family now furnishes labor and (by increased consumption) income to the economy that is destroying it is seen simply as an improvement. Thus the abstract and extremely tentative value of money is thoughtlessly allowed to replace the particular and fundamental values of the lives of household and community.[21]

It is remarkable how the commitment to perpetual economic growth has gained consensus across a wide spectrum of political positions on both sides of the established left–right divide. One reason is that economic growth largely functions as an empty screen onto which everyone can project their own dreams and hopes of a better future. For someone on the traditional right, growth may be taken to equate more wealth, more jobs, more innovation and more money to spend and invest. For someone on the traditional left, growth may be taken to mean less unemployment, more public money for welfare spending, more money for education and more opportunities for social mobility. The established right and the established left may disagree as to whether the market or the state provides the best solutions to various societal challenges but, since both market and state must be fuelled by economic growth in order to generate profits and tax revenue, both positions are tied into a commitment to economic growth. What is overlooked in this false choice between market and state are all the things that fall into the domain of neither one nor the other.

Lacan has managed to make himself unpopular among many feminists with the following slogan: 'The Woman doesn't exist.'[22] While some choose to take this slogan as an expression of some kind of sexism, I think it may provide the basis for a new critical potential for women in contemporary growth capitalism. From the perspective of eco-analysis, we can read Lacan's slogan in the light of the division of labour between Adam and Eve. 'Woman doesn't exist' thus comes to refer to the fact that the kind of maternal labour that is epitomized in giving birth to a child is not registered by the symbolic order of the market. The labour resulting in the birth of a child is 'priceless'. In our discussion of the physiocrats, we have seen how the original production of agricultural produce, *blé*, is conceived as a gift from Mother Nature and thus not immediately subsumed by the law of equivalent exchange. This is also the gist of contemporary ecological economics, which may in some respects be regarded as a revival of physiocrat economics.[23] The contribution of natural ecosystems to economic production is invisible from the perspective of conventional economics in general and neo-classical economics in particular. A Lacanian summary of ecological economics could be: 'Mother Nature doesn't exist.' The point in ecological economics is obviously not that, since she doesn't exist from the perspective of neo-classical economics, Mother Nature is in any way inferior or unimportant. In fact, it is the complete opposite.

A similar argument can be made with respect to maternal labour. Even though this kind of labour is not recorded and not priced by the market economy, it shouldn't be regarded as inferior or unimportant. Ecological economics calls for the preservation and even regeneration of the productive and reproductive capacities of natural ecosystems. In similar fashion, a truly subversive feminism would call for the preservation and regeneration of the productive and reproductive capacities of the 'household economy' that is not organized through the law of equivalent exchange of the market but through the law of impossible exchange of the gift economy. The strategy of such feminism would follow the previously quoted dictum of Žižek that 'sometimes, at least – the truly subversive thing is not to disregard the explicit letter of Law on behalf of the underlying fantasies, but to stick to this letter against the fantasy which sustains it'.[24] Along these lines, the call for perpetual GDP growth should be supported only with the provision that we take it literally as a growth in the gross *domestic* product. Perhaps this is what Fraser is aiming at:

> [T]he crisis of neoliberalism offers the chance to break the spurious link between our critique of the family wage and flexible capitalism. Reclaiming our critique of androcentrism, feminists might militate for a form of life that decenters waged work and valorizes uncommodified activities, including, but not only, carework. Now performed largely by women, such activities should become valued components of a good life for everyone.[25]

At this stage, we should be careful to not just elevate the gift into some more original, authentic, righteous or true form of human interaction. This would be at odds with Žižek's conception of the real and his insistence that it is inherently elusive and ambivalent. We should be careful not to romanticize the giving of gifts. While money may indeed institute a violent interruption of non-monetary forms of interaction, the giving of a gift may itself also contain an element of violence. This is one of the basic points of Marcel Mauss's seminal work on *The Gift*, where he sets out to answer the question: 'What power resides in the object given that causes its recipient to pay it back?'[26] The exchange of gifts between pre-monetary communities functions to establish strong bonds of reciprocity. These may be bonds of solidarity and coherence but they may also be relations of fierce competition and antagonism. The argument is that gifts carry with them the implicit obligation for reciprocation. A gift may be priceless but that does not mean that it is necessarily free. As an anthropologist, Mauss of course finds support for this argument in the study of different primitive, pre-monetary cultures, but we can also think through some of the examples that have already been touched upon in the preceding analysis.

Maternal labour was put forward as a pre-eminent example of labour offered as a gift. At first sight, this image may conjure up the associations of the unconditional love and altruistic care of the mother for a helpless, vulnerable child. But is not the child born into the world with the implicit obligation to return the love and affection of the mother and to act as the screen onto which the mother projects all her fantasies about herself? Is not the purpose of the child to act as the piece of the real completing the identity of the mother? Is the child free to choose not to love the mother? And what about our imaginary example of Rita and Bent? What if Rita never liked Bent in the first place but, when he clears the snow from her pavement, she feels obliged to invite him in for dinner? What if Bent's motivation for clearing the snow in the first place was nothing but a calculated move to make Rita offer him dinner? Another example mentioned in the above is sex, which we like to think of as a gift that is simultaneously offered in the mutual relationship between lovers. But at the same time, the giving and withholding of the gift of sex may also be an extremely efficient way for one party in a relationship to exercise power over the other party. Once you have received the gift of sex from one person, you may have implicitly accepted not to receive such a gift from another person, in so far as the receipt of the gift of sex carries with it the commitment to a monogamous relationship.

Thinking of the gift in terms of Žižek's concept of the real, the point is that the gift tends to 'resist symbolization'. Contrary to the exchange of commodities for money, the offering of a gift does not come with an explicit demand for the counter-exchange of a service or commodity of an equivalent value. The gift is priceless. This is why the offering of a gift is simultaneously an act of generosity and violence. The gift demands nothing from the recipient. But, in turn, this also means that there is nothing that the recipient may offer in return in order to equalize the relationship with the giver. If Rita does not want to enter into an affective relationship with Bent, she might prefer to pay him money for the clearing of her pavement. And, along similar lines, some people may prefer to buy sex from a prostitute rather than getting it as a gift from a lover because they do not want to become entangled in a relation of unarticulated demands and obligations. The fact that the gift can be both generous and violent is often lost in the standard feminist argument that, as long as women do not have a job to earn their own money, they are wholly dependent on their husbands and thus subject to their authority. Such an argument completely overlooks the power and violence that may be exercised through a privileged position in the gift economy of the household.

In his reading of Lacan, Žižek extends the logic of the gift to apply to the symbolic order as such:

> [L]anguage is a gift as dangerous to humanity as the horse was to the Trojans: it offers itself to our use free of charge, but once we accept it, it colonizes us. The symbolic order emerges from a gift, an offering, that marks its content as neutral in order to pose as a gift: when a gift is offered, what matters is not its content but the link between giver and receiver established when the receiver accepts the gift.[27]

This argument corresponds to the notion of symbolic castration that we have already explored. Language may immediately seem to be nothing but a smart technology that provides us with the means to express our inner thoughts and desires. The catch is of course that, once integrated into the symbolic order of language, we are caught in the logic of representation, identity and reflexivity that functions to structure our desires and prevents us from ever achieving their full satisfaction. Along these lines, we can think of the whole of the money system as a gift. This is in fact the way that Smith presents it and it is the understanding still pervading neo-classical economics. Money is nothing but a smart technology to overcome the 'clogged and embarrassed' operations of barter. The money system is 'an offering that marks its content as neutral'. Money is just a medium of exchange. But once accepted, once communities have been reorganized according to the principles of monetary exchange, we are 'colonized' by the gift of money. The paradox at play here is that by offering itself as a gift, money functions to reconfigure the whole field of exchange, thereby effacing pre-monetary forms of interaction such as gift giving. Nomy kills the eco, money kills the gift.

The call for green jobs to solve the double crisis of unemployment and environmental degradation is the symptom of this self-perpetuating dynamic in growth capitalism. The dismantling of pre-monetary forms of organization and interaction creates a demand for private corporations or government institutions to perform the functions previously filled by sharing and gift giving. On the one hand, this is itself a source of growth. On the other hand, it also creates a demand for growth as money is now needed to pay for the services provided by the market or the state. And as more and more domains of life are subsumed under the monetary logic of the market, it is becoming increasingly difficult to imagine solutions to societal challenges that are not mediated through the economy, whether it is organized by private corporations or government institutions. The narrow understanding of the economy found in neo-classical thinking tends to see the gift either as an anachronism reminiscent of previous forms of economic organization or as a superfluous luxury that may be afforded as wealth is accumulated through 'ordinary' forms of exchange. The gift figures as a liability rather than an asset, as consumption rather than production. Again, the parallel with ecology is illuminating. Natural ecosystems are crucial for the production of commodities and services within the framework of the economy. Neo-classical economics fails to register and recognize this dependence. In similar fashion, the production and exchange of commodities and services within the framework of the economy is dependent on the offering of labour as a gift. Rather than calling for more private or public sector green jobs, we should perhaps try and organize the economy in a way that would allow more people to live a life where they could offer their labour as a gift. Such revaluation of labour as gift would in turn short-circuit the ideology of perpetual growth as it might provide alternative ways of meeting the demands that are ordinarily posed to justify the need for growth. What is required is a break with the contemporary monoculture of salaried labour, where full membership of society seems to be achieved only when the subject occupies a job. Today, only labour that is priced and exchanged for money is regarded as a valuable contribution to society. Today, the big Other desires only those who desire a job.

8

The Drive for Growth

From Desire to Drive

> Therein lies the difference between desire and drive: desire is grounded in its constitutive lack, while the drive circulates around a hole, a gap in the order of being. In other words, the circular movement of the drive obeys the weird logic of the curved space in which the shortest distance between two points is not a straight line, but a curve: the drive 'knows' that the quickest way to realize its aim is to circulate around its goal-object. At the immediate level of addressing individuals, capitalism of course interpellates them as consumers, as subjects of desire, soliciting in them ever new perverse and excessive desires (for which it offers products to satisfy them); furthermore, it obviously also manipulates the 'desire to desire,' celebrating the very desire to desire ever new objects and modes of pleasure. However, even if it already manipulates desire in a way which takes into account the fact that the most elementary desire is the desire to reproduce itself as desire (and not to find satisfaction), at this level, we do not yet reach the drive. The drive inheres in capitalism at a more fundamental, systemic, level: the drive is that which propels forward the entire capitalist machinery, it is the impersonal compulsion to engage in the endless circular movement of expanded self-reproduction. We enter the mode of the drive the moment the circulation of money as capital becomes an end in itself, since the expansion of value takes place only within this constantly renewed movement.[1]

In the preceding two chapters, we have looked into need and desire respectively in relation to growth. We have seen how the necessity of growth is justified with reference to the needs of hungry poor people. We have also seen how the desire of the economic subject is structured through the division of labour. Economic castration shapes the desire for goods and services in the consumer market, the desire for jobs in the labour market, as well as the desire for money itself. In the current chapter, we shall turn to the curious concept of drive. Out of the three psychoanalytic concepts of need, desire and drive, the latter is perhaps the one that best captures the distinct self-propelling nature of contemporary growth capitalism.

As illustrated by the above quotation, the concept of drive is typically defined in relation to desire. Desire is directed at an object, which is at the same time over-determined by a series of phantasmatic projections. This is why, as Žižek points out, 'the subject... and the object-cause of its desire... are strictly correlative'.[2] It is obvious that the concept of desire is immediately applicable to consumer capitalism. This is what we have already explored on several occasions throughout the book. While the concept of desire certainly points to important dimensions of the ideology of growth, the distinctive characteristic of our current predicament perhaps lies rather in the domain of drive. 'Once we move beyond desire – that is to say, beyond the fantasy which sustains desire – we enter the strange domain of *drive*: the domain of the closed circular palpitation which finds satisfaction in endlessly repeating the same failed gesture.'[3]

The field of tension between the desire of the subject and the object of desire is sustained by fantasy. In the domain of drive, there is no such phantasmatic support. There is no belief in any form of satisfaction or relief brought about by the achievement of an object. In fact, the drive is not even oriented towards an object. While desire has a goal, which is the appropriation of the object, drive has an aim, which is the very process propelled by aim itself. The rhythm of drive is the failed repetition.

The move from desire to drive is exemplified in the development of addictions such as compulsive gambling, drug addiction and compulsive shopping.[4] Let's make a small detour to illustrate the concept of drive: the subjectivity of the ordinary, non-compulsive gambler is largely structured by the desire for money. The goal of the gambler is to win. This desire is supported by various fantasies projected onto the sublime object of money or simply the event of winning, which functions as an 'answer of the real', somehow redeeming the subjectivity of the gambler.[5] But being exposed to the groundless chance of the game also carries the risk that the desire for money and the fantasies sustaining this desire are eroded. The oscillation between winning and losing, with no apparent reason for one or the other, challenges the belief that money carries any meaning at all. In gambling, money is de-sublimated. This is the point where the gambler moves from the domain of desire into the domain of drive.[6] The goal of gambling is no longer to win money. The aim of gambling is to gamble. The gambler is submerged in the purely repetitive rhythm of the game, where no win or loss can be big enough to provide a logical conclusion to the game. This is expressed in the words of the legendary gambler, Nick 'the Greek' Dandalos: 'The next best thing to gambling and winning is gambling and losing.'[7] It is when the gambler becomes completely caught up in the mode of the drive that the state of addiction evolves. In this mode, the gambler 'finds satisfaction in endlessly repeating the same failed gesture', which carries with it all the social, emotional and economic costs associated with compulsive gambling.

Transposing the distinction between desire and drive into the domain of eco-analysis, we find a similar duality in the functioning of growth capitalism. Žižek himself already provides the initial elements of such analysis: 'At the immediate level of addressing individuals, capitalism of course interpellates them as consumers, as subjects of desire.' Economic growth is propelled by the consumer demand for ever more and ever newer goods and services. And this consumer demand is itself the result of the way that capitalism functions to structure the desire of the subject. This point should of course be expanded to also include the structuring of the desire for jobs, the desire for money and the objective desire for more commodities inherent in commodities themselves. In the final part of the introductory quotation, Žižek accounts for the move into the domain of drive. While desire is expressed at the level of the individual subject, 'drive inheres in capitalism at a more fundamental, systemic, level'. But how and where may we observe this drive at work? Unfortunately, Žižek provides little more than knee-jerk classic Marxism as he points to 'the moment the circulation of money as capital becomes an end in itself'. While this statement is so generic that it is almost impossible to disagree with, I think it still needs to be revised in order to fully capture the particular drive of contemporary growth capitalism. This paraphrase thus serves as the claim to be explored throughout the remaining part of this chapter: 'We enter the mode of the drive the moment the circulation of money as *debt* becomes an end in itself since the expansion of value takes place only within this constantly renewed movement.'

Decoupling and Post-Credit Money

> Decoupling at its simplest is reducing the amount of resources such as water or fossil fuels used to produce economic growth and delinking

economic development from environmental deterioration. For it is clear in a world of nearly seven billion people, climbing to around nine billion in 40 years' time, that growth is needed to lift people out of poverty and to generate employment for the soon to be two billion people either unemployed or underemployed.[8]

The idea of decoupling is a cornerstone in the ideology of green growth as it allows for continued economic growth without adverse ecological effects. The OECD sums up the definition of decoupling simply as breaking the link between 'environmental bads' and 'economic goods'.[9] The above quote from the United Nations provides a very condensed summary of several points from the preceding analyses as it refers to the needs of poor people, the demand for employment and the sheer growth of people on the planet. From an eco-analytical perspective, decoupling is yet another example of the ideological trend that Žižek maps out in the following:

> On today's market, we find a whole series of products deprived of their malignant property: coffee without caffeine, cream without fat, beer without alcohol...And the list goes on: what about virtual sex as sex without sex, the Colin Powell doctrine of warfare with no casualties (on our side, of course) as warfare without warfare, the contemporary redefinition of politics as the art of expert administration as politics without politics, up to today's tolerant liberal multiculturalism as an experience of Other deprived of its Otherness (the idealized Other who dances fascinating dances and has an ecologically sound holistic approach to reality, while features like wife beating remain out of sight)? Virtual Reality simply generalizes this procedure of offering a product deprived of its substance: it provides reality itself deprived of its substance, of the hard resistant kernel of the Real – just as decaffeinated coffee smells and tastes like the real coffee without being the real one, Virtual Reality is experienced as reality without being so.[10]

The purpose of products such as 'coffee without caffeine, cream without fat, beer without alcohol' is to provide consumers with an opportunity to live a more healthy life without having to change their basic habits of consumption. Instead of working to transform excessive or even addictive forms of consumption, consumers may keep on eating and drinking as much as they please. They do not have to engage with the fundamental structuring of their desires. In similar fashion, decoupling allows us to enjoy economic growth without ecological resource depletion, the reality of consumption deprived of the 'hard resistant kernel of the Real' of waste. We do not have to change the fundamental organization of the economy that is based on perpetual growth. We only need to find smarter and greener ways to produce stuff.

The standard critique of decoupling, as already mentioned, is that, even if it is in theory a good idea, it is not possible in practice.[11] A number of countries, including Germany and Japan, have managed to achieve relative uncoupling, where material resource consumption grows at a lower rate than economic growth but this is still a far cry from the kind of absolute decoupling required to seriously address the challenges of resource scarcity and climate change.[12] Critiques of decoupling which maintain that it is impossible usually draw on insights from the field of ecological economics. A standard reference is Nicholas Georgescu-Roegen, who transposes the physical laws of thermodynamics onto the field of economics.[13] The argument is that economic processes of value creation and consumption necessarily involve the use of energy and the transformation of highly ordered and thus highly useful physical matter into less ordered and less useful physical matter. This increase in disorder is also referred to as an increase in entropy.

There is an obvious intellectual lineage between contemporary ecological economics and physiocrat economics.[14] We may even refer to ecological economics as a form of neo-physiocrat economics. In the original physiocrat economics, as well as in contemporary forms of neo-physiocrat economics (ecological economics, bioeconomics, thermoeconomics, etc.), we find a conception of economic growth that puts an emphasis on the real dimension of the eco. For the physiocrats, value is ultimately the physical matter referred to as *blé*. For ecological economics or bioeconomics, the process of production and consumption ultimately amounts to using energy and transforming matter. In philosophical terms, ecological economics rejects the idea of a decoupling of the reality of economy from the real of the eco. '[M]atter, per se, also matters', as Georgescu-Roegen famously quips.[15] This is why decoupling is impossible.

At this stage, it is again crucial to maintain that the pertinent eco-analytical question is not whether decoupling is practically possible or not. This is the kind of 'wrong' question that we discussed previously. Therefore, in order to unveil the eco-analytical intricacies of decoupling, we need to reverse the usual critique of decoupling. Rather than arguing how decoupling might be a good idea in theory but impossible to realize in practice, we should invoke the French proverb: 'It's all very well in practice, but it will never work in theory.' The purpose of the fantasy of decoupling is to save the fantasy that infinite economic growth is possible. This, however, distracts us from asking why we must have economic growth in the first place. In the following, we shall see how the call for perpetual economic growth is connected to our contemporary monetary system where money is created out of interest-bearing debt by private commercial banks. The maintenance of this system depends at least on the illusion that perpetual economic growth is possible. As long as the fantasy of decoupling stands in the way of questioning this system because it supports the idea that perpetual economic growth is possible, we should reject this fantasy. Even if absolute decoupling were indeed possible, we should still push for a reorganization of the economy that curbs the imperative of perpetual growth.

The evolution and the functioning of our current system of money is the subject of my previous work on *Making Money – The Philosophy of Crisis Capitalism*. The following is a brief summary of the main argument of this book: the evolution of our current paradigm of money may crudely be conceived as an overlapping succession of two stages of decoupling. In the first stage, the creation of fiat money by the state is decoupled from the supply of precious metals such as gold and silver otherwise restricting the issuance of new money. Invoking Žižek's conception of the intricate relation between the real and the symbolic, we should be careful not to romanticize previous monetary regimes based on some form of gold standard. Even if a monetary system involves a formal gold standard, this does not necessarily mean that the money issued represents some kind of sublime value. A gold standard is ultimately nothing but a law which may be suspended or relaxed as the lawmaker sees fit. And even when the law is in place, the desire for gold is itself not independent of the law. Under a gold standard, money seems to derive its value from a reference to an underlying reserve of inherently valuable gold. But, at the same time, the seemingly inherent value of gold is nothing but a phantasmatic projection.[16] Still, the existence of a gold standard puts some form of restriction on the capacity of the money-issuing authority to expand the nominal amount of circulating currency. The nominal value of the money supply is, in other words, coupled to the material supply of gold or silver.

The history of money from 1844, when the Bank of England was granted the monopoly to print money while at the same time being committed to a gold standard, until the so-called Nixon Shock of 1971, when the United States suspended the convertibility of the US dollar into gold, thus effectively terminating the Bretton Woods Agreement, may be conceived as the gradual decoupling of central bank fiat money from an underlying reserve of material assets. Such an account is of course a gross historical and philosophical simplification and we should certainly not overestimate the importance of the event of the Nixon Shock itself. Rather, it was the inevitable outcome of tensions built up in the global monetary system. The period before and after the First World War was a time of monetary turmoil, with key currencies such as the British pound, the French franc and various versions of the German mark going on and off the gold standard. At the end of the Second World War, the US dollar was the only currency strong enough to sustain a gold standard. As a consequence, the 1944 Bretton Woods Agreement organized the major western currencies plus the Japanese Yen in a regime of fixed exchange rates around the dollar, which was in turn tied to gold. However, as the global economy and the volume of international trade grew in the decades after the Second World War, the US money supply expanded to a point where the gold standard could no longer stand the pressure from other central banks wanting to convert reserves of dollars into gold or from private investors seeking to exploit arbitrage opportunities between the open gold-market price and the fixed price guaranteed by the US Treasury. This was the background to the collapse of the US gold standard.[17]

But in order to understand our current paradigm of money, we need to include another stage of decoupling, which is the decoupling of commercial bank credit money from central bank fiat money. This is also a process that has been evolving over decades and even centuries but it has particularly picked up speed since the 1970s. Today, only a fraction of the total supply of money in western as well as other comparable economies is constituted by fiat money cash originating from state-controlled central banks.[18] The vast majority of money in circulation is commercial bank credit money.[19] This is the money sitting in commercial bank deposit accounts that we use whenever we make payments by credit cards, debit cards, bank transfers, cheques or other forms of non-cash money transactions. From an individual money-user perspective, it is easy to think of bank credit money as merely a convenient substitute for central bank physical cash. When buying things in a shop, it seems to make little difference if you pay by cash or card. But shifting into a macro perspective on the economy, there are crucial and consequential differences between the two forms of money.

While physical cash is created as the central bank runs its printing press, commercial bank credit money is created when the bank issues new credit. When a customer takes out a loan of say £100,000 from a bank, this money is typically not paid out in cash but merely credited to the customer's account in the bank. In this way, the bank essentially creates new money by simply expanding its balance sheet. On the one hand, the bank records the loan of £100,000 as a debt that the customer now owes the bank. This debt figures as an asset on the bank's balance sheet. On the other hand, the £100,000 that is paid out to the customer is recorded as a credit against the bank. The customer's deposit is money that is owed by the bank. This credit figures as a liability on the bank's balance sheet. While the books of the bank still balance, it has nevertheless managed to create £100,000 worth of new money out of nothing. Or perhaps more correctly, it has created £100,000 worth of new money out of the customer's debt to the bank. The gist of the matter is of course that in our current economy, where most payments are made as bank credit transfers, credit against a bank *is* money. It is thus misleading to think of banks as merely 'financial intermediaries'. Rather than lending out money already in existence, banks are capable of making money by the same token as they lend it out.[20] This is how the majority of money circulating in our contemporary economy has come into being.

It is true that when customers start spending this money and credit flows back and forth between different banks, the banks need to hold a certain amount of reserves in the form of central bank credits in order to clear outstanding balances with other banks. It is also true that financial authorities have set up certain capital requirements that banks are legally obliged to observe. But in practice these requirements put only very limited restrictions on the capacity of banks to create new money. Since the 1980s, we have seen a shift in monetary policy whereby central banks no longer see their role as limiting the credit creation of commercial banks but rather just aiding banks in their capacity to meet whatever demand for credit exists in the economy.[21] During the unfolding of the financial crisis of 2007–8, we saw how central banks willingly stepped in to supply additional liquidity in the form of central bank credits as the inter-bank money market dried up. Furthermore, the central banks of the United States, the United Kingdom and even the European Union have engaged in several rounds of quantitative easing, where even more central bank credit is injected into the money markets in an attempt to entice commercial banks to extend their credit to productive enterprises.

The combination of these two forms of decoupling constitutes what I refer to as the paradigm of post-credit money:

> The dominant form of money today is post-credit money. This kind of money is created by private commercial banks as they issue credit circulating in an extensive and efficient credit payment system.... The issuance and circulation of private bank credit money is supported by states and the central banks through a number of measures. First of

all, the state has allowed and endorsed the emergence of the electronic credit payments system thus outsourcing control over the monetary infrastructure of the economy to private agents with special privileges.... Secondly, central banks throughout the Western world have adopted policies to intervene in interbank money markets and other sectors of the private banking system only to support the maintenance of this system.... And finally, even though fiat cash money constitutes only a fraction of all money in circulation, it still plays the crucial ideological role of providing the ideal fantasmatic image of money as such.[22]

In each of these two forms of decoupling constituting the emergence of post-credit money, we see a similar logic as the one pointed out by Žižek in 'coffee without caffeine, cream without fat, beer without alcohol'. The decoupling of money from gold allows for the supply of fiat money to expand without any depletion of gold reserves. And the decoupling of commercial bank credit money from central bank fiat money allows for the supply of credit money to expand without being restricted by the provision of central bank money reserves.

The system of post-credit money rests on the fantasy that virtual credit money and physical cash money are ultimately exchangeable at par. The ATM is an ideological apparatus that supports this fantasy by offering cash money in exchange for credit money whenever the customer should want this.[23] As long as only a fraction of customers take advantage of this offer for only a fraction of their deposits, the fantasy successfully creates its own conditions of possibility. And as long as the fantasy is operative, there is no actual need for people to take out their credit money in cash. The decoupling of credit money from fiat money is intact. But the excessive expansion of commercial bank balance sheets ultimately puts a strain on the convertibility of credit money into cash at par, that is, the ability to convert a bank deposit of 1,000 euros into 1,000 euros in cash. If bank deposit holders lose faith in the bank's capacity to recover its outstanding loans, they may want to secure the value of their deposits by taking them out of the bank. The bank is now faced with the risk of a bank run and the bank customers are faced with the risk of a bail-in, where their assets in the bank are seized in order to cover the bank's losses on defaulting creditors. In philosophical terms, the breakdown of par signifies the collapse of the fantasy that veils the traumatic split between credit money and fiat money, with the result that the real re-emerges as the 'impossible/traumatic kernel' or the 'central lack' around which the symbolic order of money is structured.[24] In economic terms, the breakdown of par signifies inflation in credit money and deflation in cash money as depositors 'flee to security'.

The Parable of the Eleventh Round

We have already touched upon various dimensions of the relation between money and growth throughout the preceding analysis. The purpose of the current as well as the following sections is to explore how the relation between money and growth is reconfigured with the emergence of post-credit money. In the classical account of money as a means of exchange to alleviate the inconveniences of barter, money is a mere facilitator of growth because it improves the efficiency of the market. Money enables the exchange of diverse commodities and services, thus liberating the butcher to concentrate on butchering, the baker to concentrate on baking and the brewer to concentrate on brewing. In other words, money facilitates the division of labour, which in turn creates growth through increased productivity. In our critique of this conception of money, we have seen how the relation between money and growth is much more intricate. Money not only facilitates growth by enabling productivity, although, admittedly, it does that too. Money plays a key role in the very constitution of the desires of the economic subject, thus creating new and increasing demands for new commodities and services that can only be provided through the market economy. In addition, money itself emerges as an object of desire, thus furthering the expansive forces of the market economy. In other words, money not only responds to pre-existing needs that are served through the market economy. Through the structuring of desire, money puts into motion a self-perpetuating movement towards more and more growth.

A comparable insight into the dialectical relation between money and growth is at the heart of Keynesian economics. In times of recessive growth, the state may step in to expand the supply of money, thus sparking the economy's inherent propensity to grow. This is an operational application of the Lacanian dictum that desire is the desire of the big Other. As the state provides a stimulus to the economy through government spending, it acts as the big Other, desiring the labour, commodities and services otherwise not demanded in the economy. The effects of this incarnation of the big Other by the state multiplies as the desire of all other agents in the economy is subsequently also stimulated. The initial demand by the state for the private agents is translated into mutual demands among the private agents in the economy. The conventional wisdom of mainstream economics in general and the so-called quantity theory of money in particular is that, if the supply of money grows faster than the underlying economy, the price of money decreases and inflation sets in. On a strict reading, the quantity theory of money would therefore not allow for such stimulus to have any effect on growth since the government expansion of the money supply would simply just result in inflation in the form of an increase in nominal prices. While Keynes does indeed agree with the general principle of the quantity theory, he adds the point that in practice there is a time lag before prices react to an expansion of the money supply. This is also known as price stickiness, which allows for the level of economic activity to increase before the effects of inflation kick in.

While both of these relations between money and growth still apply today, the system of post-credit money adds another dimension to the dialectic. In our contemporary system, the creation of money is intertwined with the creation of interest-bearing debt. On the one hand, credit money itself is nothing but the debt of the bank to the bank customer. On the other hand, credit money is created as the contra-entry to the creation of a debt of the customer to the bank. On the balance sheet of the bank, the bank's debt to the customer figures as a liability and the customer's debt to the bank figures as an asset. From this immediate perspective, the two entries, money and debt, appear to be symmetrical. But if we add the dimension of temporality, the symmetry disappears. The customer's debt to the bank earns an interest depending on the conditions of the loan and the credibility of the customer. An asset backed mortgage may earn as little as 2–3 per cent annually, while a credit card loan may earn up to 20–30 per cent or even more. The bank's debt to the customer, however, typically earns very little if any interest at all. This is because this debt functions at the same time as liquid money.

If we believe in the textbook theory of banking, debt and interest, where the issuance of a loan is merely the movement of money from one agent, the lender, to another agent, the borrower, the interest differential between debt and money makes a lot of sense. The interest paid by the borrower is a compensation to the lender because he forgoes the ability to dispose of the money otherwise. This is similar to the way that a person lending out his car might expect some form of compensation, as he cannot use the car himself while it is being lent out. But, as we have seen, this is not how banks function in the age of post-credit money. Of course, a bank requires initial investors in order to come into being. But once the bank is up and running and minimal requirements for capital and liquidity are fulfilled, it does not have to wait for depositors to lend out money. It is capable of creating money through the expansion of its own balance sheet by the same token that it lends out this money. In fact, we can think of the borrower as being at the same time a lender in so far as the money paid out through the loan is nothing but a deposit with the bank. Banks transform debt into money.[25] In this light, interest seems less justified by the argument that for every debt there is a creditor, who has forgone the opportunity to dispose of this money. In the case of banks, the money would not be there in the first place if it wasn't for the debtor being willing to borrow it.

The issue at hand is not so much the moral arguments for or against interest. The point is rather that, in our current monetary system, debts tend to grow at a faster pace than the supply of money. This is due, primarily, to the basic differential between interest on debt and interest on deposits. From the perspective of the individual debtor, this is not necessarily a problem. If he has a steady job, a profitable business or another source of income, he may be perfectly capable of recovering the amount of money that is needed to make the required payments of interest and principal as they come due. But from the perspective of the total economy, the difference between the growth rate of debts and the growth rate of money amounts to a structural lack of money. There is simply not enough money to go round to meet the payments of debt and principal. This configuration of the money system has implications for the functioning of the economy. Money no longer just structures a desire for economic growth. The build-up of debt in the system of post-credit money propels the economy into a drive towards perpetual debt.

To see how this works, we shall start by looking at a parable of money and debt told by Berhard Lietaer, who by the way is one of the architects of the euro.

> Once upon a time, in a small village in the Outback, people used barter for all their transactions. On every market day, people walked around with chickens, eggs, hams, and breads, and engaged in prolonged negotiations among themselves to exchange what they needed. At key periods of the year, like harvests or whenever someone's barn needed big repairs after a storm, people recalled the tradition of helping each other out that they had brought from the old country. They knew that if they had a problem someday, others would aid them in return.
>
> One market day, a stranger with shiny black shoes and an elegant white hat came by and observed the whole process with a sardonic smile. When he saw one farmer running around to corral the six chickens he wanted to exchange for a big ham, he could not refrain from laughing. 'Poor people,' he said, 'so primitive.' The farmer's wife

overheard him and challenged the stranger, 'Do you think you can do a better job handling chickens?' 'Chickens, no,' responded the stranger. 'But there is a much better way to eliminate all that hassle.' 'Oh yes, how so?' asked the woman. 'See that tree there?' the stranger replied. 'Well, I will go wait there for one of you to bring me one large cowhide. Then have every family visit me. I'll explain the better way.'

And so it happened. He took the cowhide, and cut perfect leather rounds in it, and put an elaborate and graceful little stamp on each round. Then he gave to each family 10 rounds, and explained that each represented the value of one chicken. 'Now you can trade and bargain with the rounds instead of the unwieldy chickens,' he explained.

It made sense. Everybody was impressed with the man with the shiny shoes and inspiring hat.

'Oh, by the way,' he added after every family had received their 10 rounds, 'in a year's time, I will come back and sit under that same tree. I want you to each bring me back 11 rounds. That 11th round is a token of appreciation for the technological improvement I just made possible in your lives.' 'But where will the 11th round come from?' asked the farmer with the six chickens. 'You'll see,' said the man with a reassuring smile.

Assuming that the population and its annual production remain exactly the same during that next year, what do you think had to happen? Remember, that 11th round was never created. Therefore, bottom line, one of each 11 families will have to lose all its rounds, even if everybody managed their affairs well, in order to provide the 11th round to 10 others.

So when a storm threatened the crop of one of the families, people became less generous with their time to help bring it in before disaster struck. While it was much more convenient to exchange the rounds instead of the chickens on market days, the new game also had the unintended side effect of actively discouraging the spontaneous cooperation that was traditional in the village. Instead, the new money game was generating a systemic undertow of competition among all the participants.[26]

Since there are not enough rounds to pay back the stranger, the families now have to compete against each other. Some individual families will probably do fine under this system. The most competitive families will have no problems earning enough rounds to pay back the stranger. But some other families will eventually not have enough rounds to pay him. What the stranger has done is to simply create money on the back of interest-bearing debt. Over the course of the year, the amount of debt has outgrown the amount of money available for repayment. The scarcity of money forces an environment of competition rather than cooperation.

Initially, Lietaer merely uses the parable to illustrate how the institution of debt tends to erode non-monetary forms of economic interaction such as sharing and gift giving. Eisenstein, however, expands on Lietaer's parable and shows how the competition for rounds to pay back the debts to the stranger ultimately forces the economy of the village to grow. Unless the villagers choose to unite and kick the stranger out of the village, the economy of the village is, according to Eisenstein, likely to evolve into a situation of default and perpetual debt creation. In order to compensate for the lack of rounds to pay back both principal and interest, the stranger agrees to create new rounds and lend them out to the village families on the condition that these new rounds are also paid back with interest:

> So imagine now that the villagers gather round the man in the hat and say, 'Sir, could you please give us some additional rounds so that none of us need go bankrupt?'
> The man says, 'I will, but only to those who can assure me they will pay me back. Since each round is worth one chicken, I'll lend new rounds to people who have more chickens than the number of rounds they already owe me. That way, if they don't pay back the rounds, I can seize their chickens instead. Oh, and because I'm such a nice guy, I'll even create new rounds for people who don't have additional chickens right now, if they can persuade me that they will breed more chickens in the future. So show me your business plan! Show me that you are trustworthy (one villager can create "credit reports" to help you do that). I'll lend at 10 percent – if you are a clever breeder, you can increase your flock by 20 percent per year, pay me back, and get rich yourself, too.'[27]

Since the stranger only lends to families that are able to put up chickens or other assets in collateral for the loan or to families able to present a viable business plan for how to produce more chickens in the future, the families are compelled to increase their production of chickens in order to remain creditworthy in the eyes of the stranger, rather than to satisfy any actual demand for chickens. The financing of the economy of the village through the creation of money out of debt creates an imperative of perpetual growth.

The Imbalance Sheet of the Stranger

While the general reasoning of Eisenstein is indeed convincing, there are a number of implicit steps and assumptions in his argument that are worth examining in more detail. One of the classic tricks of mainstream economics is to begin with a very simple model that has a number of completely unrealistic assumptions and then to deduce a wide range of general economic principles and correlations from the model, which are then more or less uncritically applied to the actual economy. In the following, we are going to do just that. We shall be taking the parable of the eleventh round as the starting point for an exploration of the relation between economic growth and the creation of money out of interest-bearing debt. As a first step in this exploration, we shall apply the method of accounting to analyse the economy of the village. We assume that the village comprises a total of 10 families. Each family owns a stock of capital consisting of buildings, farming equipment, and animals that is worth the equivalent of 1,000 chickens. We shall be following the economy of three families: the Ericsens, the Jensens and the Sorensens. As a method of tracking the development of the economy of the villagers, we shall apply the principle of double-entry bookkeeping. The balance sheets of the three families thus look like this:

Year Zero

Ericsen Family

Assets	Liabilities
Capital: Ø1,000	Equity: Ø1,000
Ø1,000	Ø1,000

Jensen Family

Assets	Liabilities
Capital: Ø1,000	Equity: Ø1,000
Ø1,000	Ø1,000

Sorensen Family

Assets	Liabilities
Capital: Ø1,000	Equity: Ø1,000
Ø1,000	Ø1,000

Now the stranger enters the picture and issues 100 rounds of cowhide. We assume that each round represents the value of not one but 10 chickens. Each of the 10 families borrows 10 rounds while promising to pay back the stranger 11 rounds at the end of the year. The balance sheets of the families as well as the stranger now look like this:

Primo Year One

Ericsen Family

Assets	Liabilities
Capital: Ø1,000	Debt: Ø100
Rounds: Ø100	Equity: Ø1,000
Ø1,100	Ø1,100

Jensen Family

Assets	Liabilities
Capital: Ø1,000	Debt: Ø100
Rounds: Ø100	Equity: Ø1,000
Ø1,100	Ø1,100

Sorensen Family

Assets	Liabilities
Capital: Ø1,000	Debt: Ø100
Rounds: Ø100	Equity: Ø1,000
Ø1,100	Ø1,100

Stranger

Assets	Liabilities
Loans: Ø1,000	Debt: Ø1,000
	Equity: Ø0
Ø1,000	Ø1,000

In addition to their existing capital, each family now holds 100 chickens' worth of rounds that they can use as payment when exchanging goods and services with each other. The counter-entry of this new asset is a debt to the stranger also of 100 chickens. Looking at the balance sheet of the stranger, it seems that he has merely broken even. The rounds issued to the villagers represent a debt of 1,000 chickens, while at the same time it is also a loan of 1,000 chickens that must be repaid. Since the stranger has not yet made any profits, his equity is still zero.

The recording of the entry of the stranger into the village economy provides a simple illustration of the process through which banks use their balance sheets to create money out of debt. Looking at the balance sheet of the stranger, we can recognize a structure similar to the one found in Žižek's concept of the split subject. As we have seen previously, symbolic castration means that when the subject emerges through inclusion in the symbolic order, it finds itself in the gap between the real and the symbolic. The identity of the subject is expressed through its symbolic mandate but, at the same time, the subject insists on being more than this mandate. Žižek sometimes elaborates on this dialectic between symbolic identification and alienation in the constitution of subjectivity through the following Marx Brothers' joke:

'You remind me of Emanuel Ravelli.'
'But I am Emanuel Ravelli.'
'Then no wonder you look like him!'[28]

There is in the joke an insistence on the non-identity between the symbolic mandate of the subject – the signification of the subject by the name Emanuel Ravelli – and the real presence of the person designated as Emanuel Ravelli. It is in this minimal gap between the incomplete signification and the real presence of the subject that subjectivity manifests itself. The insufficiency of the symbolic order opens a space for the subject's imaginary projections of his or her own subjectivity. It is the feeling of alienation that constitutes what is 'in you more than yourself': 'I am Emanuel Ravelli but I am also more than Emanuel Ravelli.'

The balance sheet, which is drawn up according to the principles of double-entry bookkeeping, is the symbolic representation of the stranger's 'bank'. It is a recording of the assets and liabilities of this bank. This symbolic order of the balance sheet seems to merely state the identity of the bank, that is, the identity between debit and credit. The two sides balance out at 1,000 chickens. But the symbolic recording of the identity of debit and credit also involves an element of alienation that functions to bring out what is 'in the bank more than a bank'. While the loan of 1,000 chickens that is registered on the debit side of the balance sheet is nothing but a debt of 1,000 chickens, the deposit of 1,000 chickens on the credit side of the balance sheet is not just a debt. It is a recording of the credit issued by the bank in the form of rounds. These rounds not only function as debt but simultaneously also as money. This corresponds to the way that the credit of a bank today is a liquid means of exchange, even if it circulates as digital entries on a ledger rather than as physical leather rounds. The difference between debt and money corresponds to the Lacanian concept of *objet petit a*, which 'stands for the unknown x, the noumenal core of the object beyond appearances, for what is "in you more than yourself." '[29] In this case, *objet petit a* stands for that in credit which is more than debit. We can thus imagine the following exchange between the stranger and one of the villagers pondering the alchemical way in which money seems to have been created out of a mixture of cowhide and debt:

Villager: 'That debt represented by a token of cowhide reminds me of money.'
Stranger: 'But it is money.'
Villager: 'Then no wonder it looks like money!'

As we know, the unbalanced structure of the split subject is also what propels the desire of the subject. In this sense, the imbalance is productive. The same goes for the imbalance of the bank. The introduction of money and debt into the economy also increases the productivity of the economy. This becomes obvious when we follow the development of the village. As we move forward one year to see what the balance sheets of the village economy look like when the stranger returns to reclaim repayment of the rounds he has lent out, we shall introduce a number of assumptions not present in either Lietaer's original formulation of the parable or Eisenstein's elaboration. First, it is important that we change the nominal standard of account into cowhide rounds rather than chickens. We shall designate the rounds using the following symbol: Ø. This change is important as the stranger is not going to accept chickens in payment of the debt. Even though the cowhide rounds were initially introduced at the value of 10 chickens, this rate is not fixed. The rounds are, in other words, not backed by a 'chicken standard'.

Second, we are going to assume that the introduction of the rounds into the village economy has indeed made exchange between the families easier and more efficient. Instead of spending time on the market, bartering back and forth in order to get the things they need, their new monetary system leaves them with more time to attend to their production of chickens, eggs, ham, bread and other commodities. The increased productivity means that each family has been able to increase its total stock of capital by the equivalent of 10 chickens. In other words, the GDP of the village has grown. Since the value of this additional capital has not yet been realized in the market, it is 'marked to market', which means that it is booked at existing market prices, where 10 chickens is worth Ø1. The balance sheets of the families, as well as that of the stranger, now look like this:

Ultimo Year One

Ericsen Family

Assets	Liabilities
Capital: Ø101	Debt: Ø11
Rounds: Ø10	Equity: Ø100
Ø111	Ø111

Jensen Family

Assets	Liabilities
Capital: Ø101	Debt: Ø11
Rounds: Ø10	Equity: Ø100
Ø111	Ø111

Sorensen Family

Assets	Liabilities
Capital: Ø101	Debt: Ø11
Rounds: Ø10	Equity: Ø100
Ø111	Ø111

Stranger

Assets	Liabilities
Loans: Ø110	Debt: Ø100
	Equity: Ø10
Ø110	Ø110

Even though the debts of each of the families have increased by Ø1, this is compensated by the increase in their capital, so that their equity is still Ø100. The interests incurred on the loans issued by the stranger are of course also recoded as an increase in his equity. But despite the increase in the productivity of the villagers, they are still unable to hold good on their promise to repay the stranger his 11 rounds. Fortunately, the stranger is not an unreasonable man. He makes the proposition to the villagers that they only pay him the eleventh round, while the principal debt of Ø10 is rolled over and extended for another year. Since neither of the families are able to repay all of their debt to the stranger, the villagers see no other option than to accept this proposition. In addition, a couple of the families are quite satisfied with the way that they have been able to expand their production. They have ideas of how to expand even further and they believe that a continuation of the stranger's system is necessary in order to do so. As the proposition is accepted and each of the families has used one of their rounds to pay the interest to the stranger, the village balance sheets look like this:

Primo Year Two

Ericsen Family

Assets	Liabilities
Capital: Ø101	Debt: Ø10
Rounds: Ø9	Equity: Ø100
Ø110	Ø110

Jensen Family

Assets	Liabilities
Capital: Ø101	Debt: Ø10
Rounds: Ø9	Equity: Ø100
Ø110	Ø110

Sorensen Family

Assets	Liabilities
Capital: Ø101	Debt: Ø10
Rounds: Ø9	Equity: Ø100
Ø110	Ø110

Stranger

Assets	Liabilities
Loans: Ø100	Debt: Ø90
	Equity: Ø10
Ø100	Ø100

Now we see how the equity of the stranger has miraculously increased from zero to Ø10, which is the equivalent of 100 chickens. In our eco-analytical account, the balance sheet of the bank is the symbolic order that enables the conversion of debt into money. Continuing along these lines, the notion of equity constitutes the dimension of the imaginary, which functions to manage the split between debit and credit. The balance sheet creates the split between debit and credit and the notion of equity veils this split. The increase in equity represents an increase in the total value of the bank. And yet the only value created is a net decrease in the bank's debt to the customers relative to the customers' debt to the bank. The equity is an imaginary projection that relies on a fantasy that at some point in the future the value of the claim on the customer may somehow be redeemed. This corresponds to the way that subjects project certain images onto objects of desire, thus fantasizing that the appropriation of these objects will redeem their identity and self-realization at some time in the future. The idea that the equity of the bank corresponds to real wealth is itself a fantasy. It is true that the equity of the bank represents a claim on real assets in so far as the debt owed to the bank is backed by collateral. However, this does not mean that an increase in the equity of a bank necessarily corresponds to an increase in the volume of real assets in the economy.

The following year, the stranger comes back to collect the debts of the families and the same thing happens. The families are unable to repay the rounds that they owe, so they accept the stranger's proposition to roll over the principal and only pay the interest of Ø1. And again, the next year, the routine is repeated. But five years after the introduction of the stranger's system, a new situation has emerged. Over the course of the years, the economies of the families have begun to evolve differently. While a couple of the families, including the Ericsens, have been able to expand their production and to accumulate a relatively larger portion of the rounds in circulation, other families are struggling to keep up with the competition. Now one of the families, the Sorensens, is out of its last round and thus unable to make the payment of Ø1 to the stranger. Here is what the situation looks like on the balance sheets of our three families as well as that of the stranger:

Ultimo Year Five

Ericsen Family

Assets	Liabilities
Capital: Ø120	Debt: Ø11
Rounds: Ø12	Equity: Ø121
Ø132	Ø132

Jensen Family

Assets	Liabilities
Capital: Ø101	Debt: Ø11
Rounds: Ø6	Equity: Ø97
Ø107	Ø108

Sorensen Family

Assets	Liabilities
Capital: Ø80	Debt: Ø11
Rounds: Ø0	Equity: Ø69
Ø80	Ø80

Stranger

Assets	Liabilities
Loans: Ø110	Debt: Ø60
	Equity: Ø50
Ø110	Ø110

We see here how the families are doing differently. The Ericsen family is doing very well having increased both capital and rounds. The Jensens are in a middle group of families in the village. They have not increased their capital since the first year but they are still able to make interest payments out of their stock of rounds. The Sorensens, however, are now unable to make payments to the stranger. Not only is the family out of rounds but they have also been struggling to earn enough rounds to pay for chicken food, tractor fuel, repairs and other things needed to maintain their chicken farm. Over the past couple of years, the Sorensens' stock of chickens has decreased and their buildings and equipment have gradually deteriorated. The value of the family's capital is thus down to Ø80. We also see how the equity of the stranger has gradually increased as he collects yearly payments of interests on the original debt that is perpetually rolled over.

Eco-analysis provides a further elaboration of this imbalance introduced into the economy by the stranger's creation of money out of debt. This imbalance is structurally similar to the imbalance that is instituted with the inclusion of the subject in the symbolic order. As we have seen, Žižek theorizes this inclusion as symbolic castration and he provides the following account:

> What Lacan calls 'symbolic castration' is a deprivation, a gesture of taking away (the loss of the ultimate and absolute – 'incestuous' – object of desire) which is in itself giving, productive, generative, opening up and sustaining the space of desire and of meaning. The frustrating nature of our human existence, the very fact that our lives are forever out of joint, marked by a traumatic imbalance, is what propels us towards permanent creativity.[30]

If we transpose this definition into the field of money and debt, it is at first glance much in line with the standard account of the function of banks and credit. When banks create money on the back of interest-bearing debt, a 'traumatic imbalance' is instituted in the economy. In so far as the perpetually rising debt can never be repaid, this puts the economy 'forever out of joint'. But even if this state of perpetual indebtedness is indeed 'frustrating', it is at the same time also 'productive, generative, opening up and sustaining the space of' investment, innovation and growth. Now let's return to the village to see how the dialectics between frustration and innovation plays out.

In order to resolve the situation of the Sorensens, the stranger comes up with a new proposition. In addition to rolling over the debts of those families able to pay their eleventh round, he offers to lend back to the struggling family ten of the original rounds that he has been collecting over the years. This will allow them to not only pay their eleventh round but also to buy the things needed to get their chicken farm up and running again. Just as the Sorensens are about to accept the proposition, the stranger adds another condition. In order to get the loan, the family has to promise to repay not just eleven but thirteen rounds. This condition also applies to the original debt of 10 rounds, which will now also incur an interest of 30 per cent. The family is shocked by these conditions but still they see no other option than to accept the stranger's proposition.

After one more year, the sixth year after the stranger's first visit, the economic situation in the village has become even more severe. The Sorensens, who took out the additional loan, have not managed to turn their farming business around. They have either sold or eaten their last chickens and spent their last rounds purchasing the bare necessities to feed themselves. What is left of their initial capital is their farmland, their barn, some worn-out equipment and the house they live in. In addition, two of the other families, including the Jensens, are now also out of rounds to make interest payments to the stranger. They find themselves in a similar position to that of the Sorensens one year ago. In contrast, however, one of the village families, the Ericsens, is doing better and better each year. The Ericsens are constantly coming up with new ideas about how to improve their farming and they are able to offer chickens in the market at prices lower than anyone else. Since the Sorensens have gone out of business, they have been buying most of their food from the Ericsens. This is what the situation looks like:

Ultimo Year Six

Ericsen Family

Assets	Liabilities
Capital: Ø140	Debt: Ø11
Rounds: Ø20	Equity: Ø149
Ø160	Ø160

Jensen Family

Assets	Liabilities
Capital: Ø80	Debt: Ø11
Rounds: Ø0	Equity: Ø69
Ø80	Ø80

Sorensen Family

Assets	Liabilities
Capital: Ø60	Debt: Ø26
Rounds: Ø0	Equity: Ø34
Ø60	Ø60

Stranger

Assets	Liabilities
Loans: Ø125	Debt: Ø60
	Equity: Ø65
Ø125	Ø125

When the stranger comes back for his annual visit, unrest is spreading throughout the village. The Sorensens, the Jensens and other families in the village who are struggling to make ends meet are frustrated by their own situation. The rest of the village families are sorry to see how their neighbours and friends are going from bad to worse. Still, feelings in the village are not wholly unambiguous. A couple of families, including the Ericsens, are doing better than ever and the total production of the village has increased since the stranger first came. While many families are worried about how to keep making payments to the stranger, they also experience a sense of increasing affluence. Chickens, eggs, ham and other products are becoming cheaper and cheaper and, rather than just selling these raw, some families have specialized in turning them into delicacies that are sold in special stalls at the market. As the stranger listens carefully to the concerns of the villagers, he promises to work with them to find solutions to their problems. He sits down with the Sorensens and the Ericsens and jointly makes them the following proposal: the Sorensens are now Ø26 into debt with no immediate prospect of future earnings. Furthermore, the Ericsens are obviously more successful than the Sorensens at running a chicken farm so everyone is probably better off if the Ericsens take over the Sorensens' family farm. The stranger thus offers to broker a deal where the Ericsens buy the Sorensens' farm for Ø50. This allows the Sorensens to pay off their debt and even have some spare rounds left to spend. The Ericsens will then employ the Sorensens to work on the farm as it is merged with their existing business. The Sorensens will even be able to stay in their house for a monthly rent that is paid out of their salary. Of course, the Ericsens only have Ø20 of savings so the stranger offers to lend them an extra Ø40 to finance the deal. The conditions of this loan are relatively favourable as the Ericsens will have to pay an annual interest of only 5 per cent.

With the prospect of expanding their business and also getting a good bargain on the Sorensen farm, the Ericsens readily accept their part of the deal. Eventually, the Sorensens also agree, although they are more hesitant than the Ericsens. The Jensen family and the other family unable to make their payment of the annual eleventh round to the stranger are offered a choice between borrowing an additional Ø10 with the promise of paying back Ø13 in a year or selling their farm to another one of the village families that is doing well. Seeing how the Sorensens have ended up having to sell their farm anyway, the Jensens decide to cut their losses. The Jensen wife has already started working part-time with another family, keeping their house and preparing delicacies to be sold in the market, so the change to becoming salaried workers rather than an independent farming family is not sudden. A deal is brokered whereby the Jensens get to keep their house, valued at Ø30, and only sell off their farmland, barn and equipment, at a Ø10 discount, for Ø40 to another family. This family borrows Ø40 at 5 per cent interest from the stranger to finance the deal. The other family that is unable to make their payment to the stranger decides to fight to keep their farm. They agree with the stranger that they will get a new loan of not just Ø10 but Ø20 and pay 30 per cent interest on their entire debt. When the stranger leaves the village this time, the balance sheets look like this:

Primo Year Seven

Ericsen Family

Assets	Liabilities
Capital: Ø200	Debt: Ø50
Rounds: Ø9	Equity: Ø159
Ø209	Ø209

Jensen Family

Assets	Liabilities
Capital: Ø30	Debt: Ø0
Rounds: Ø29	Equity: Ø59
Ø59	Ø59

Sorensen Family

Assets	Liabilities
Capital: Ø0	Debt: Ø0
Rounds: Ø24	Equity: Ø24
Ø24	Ø24

Stranger

Assets	Liabilities
Loans: Ø190	Debt: Ø125
	Equity: Ø65
Ø170	Ø170

We see how the village is developing into a heterogeneous economic organization, with each of the families as well as the stranger moving into different class positions. The Ericsens are evolving into capitalists, earning a profit on the productive application of capital as well as the incorporation of salaried labour. The Sorensens are now purely labourers with no capital of their own. And the Jensens retain a position somewhere in the middle as they no longer own any productive capital but still own their own house. The stranger, obviously, sits in a position within the financial class of rentiers earning money through interest on debts. In addition, we can also see how the demand for money in the village economy has evolved and become increasingly diverse. Money is needed not only as a means to exchange commodities but also for the payment of salaries, rent and interest.

Now let's imagine the following events in year seven: as the Sorensens have no means of producing their own food and also no longer own their own house, they are compelled to earn enough money to buy all their necessities in the market, as well as to pay their rent to the Ericsens. After some months, the Sorensens have to take money out of their savings to make ends meet, and they find their initial sum of Ø24 to be gradually dwindling. One day, the Sorensen father has an accident in the Ericsens' field, leaving him unable to work. This turns out to be the last straw breaking the back of the family's economy, as they now have to live off only one income. Their last rounds are soon spent on food. The Ericsens agree to lend the Sorensens money for rent until the father is able to start working again. But after three months, with no immediate recovery in sight, the line of credit runs out and the Sorensens are evicted from their house. Angry and frustrated with the situation, the Sorensen father gets into a fight with the Ericsen father. This leads to the Sorensen mother being fired from her job with the Ericsens. The Sorensens are now left on the street with no job.

Always entrepreneurial, the Ericsens now invest Ø30 of their earnings in renovating the house where the Sorensens used to live and dividing it into two separate condominiums. One condominium is sold for Ø50 to the Jensens' son, who wants to move out of his parents' house. The Jensen son receives Ø10 from his father as a gift, and he borrows the remaining Ø40 to finance the deal from the stranger at 20 per cent interest. The economy of the Jensen family is otherwise unaltered. The other condominium is sold to the daughter of another village family under similar conditions. In addition to making a gain on the renovation and sale of the two condominiums, the Ericsens have seen a growth of Ø20 in their capital, due to a general increase in their production and turnover. Still, the sale of the two condominiums means that Ø30 is subtracted from their stock of capital. After the stranger's annual visit at the end of year seven, where interest is paid and new loans issued, we can imagine that the balance sheets of the villagers look like this:

Primo Year Eight

Ericsen Family

Assets	Liabilities
Capital: Ø190	Debt: Ø50
Rounds: Ø100	Equity: Ø270
Ø320	Ø320

Jensen Family

Assets	Liabilities
Capital: Ø30	Debt: Ø0
Rounds: Ø19	Equity: Ø49
Ø59	Ø59

Sorensen Family

Assets	Liabilities
Capital: Ø0	Debt: Ø0
Rounds: Ø0	Equity: Ø0
Ø0	Ø0

Stranger

Assets	Liabilities
Loans: Ø260	Debt: Ø155
	Equity: Ø105
Ø260	Ø260

Jensen Son

Assets	Liabilities
Capital: Ø50	Debt: Ø40
Rounds: Ø0	Equity: Ø10
Ø40	Ø40

At this stage, it would be reasonable to assume that some of the other families are beginning to find it difficult, if not impossible, to pay their interest to the stranger. In particular, the family that took out an additional loan of Ø20 at 30 per cent interest will almost certainly have had to default on their debt. However, our model has grown so complex that we would need to include other assumptions to continue. I shall thus stop the simulation at this stage and leave it up to the reader to include new scenarios in the model.

Looking at the development of the imaginary village economy, we can see the contours of a number of interconnected trends that illustrate the dialectics between economic growth and the creation of money out of interest-bearing debt. First, we see how the class division of the village economy continues and there is an increase in inequality between the families as illustrated by the growing differences in their equity. Of course, the class structure of the village also includes the stranger, and we see how the equity of his 'bank' is also steadily increasing. Second, the equity of the stranger's bank is also an expression of the difference between the number of outstanding interest-bearing loans and the amount of money in circulation. The debt registered as a liability on the stranger's balance sheet is an expression of the total village supply of money. For the first five years, this difference increases as the money in circulation gradually decreases as it is used to make the annual payments of interest. The level of outstanding loans is constant as the principals are simply rolled over. But after year five, when the stranger begins to make additional loans, both loans and money supply start to increase. However, the rate at which the supply of money grows is necessarily slower than the growth of loans, since the issuance of new loans always creates additional debt as well as an additional demand for money to pay interest. Third, the equity of the stranger's bank represents his stake in the capital of the village over and above the amount of rounds that have been issued. The villagers owe the stranger more than the supply of money in circulation. This means that their capital takes on the character of collateral. As the equity of the stranger increases, the amount of capital that functions as collateral for loans increases equally. This gradual collateralization of the capital of the village stands for a financialization of the economy. The value of houses, farms and land is no longer measured merely in terms of their use-value for the villagers but also in terms of the price at which they can be bought and sold on the market.

Fourth, finally and most importantly, the discrepancy between the amount of outstanding loans and the amount of circulating money makes it impossible for the village as a whole to repay its debt to the stranger. This affords the stranger significant power over the economy of the village. The village is dependent on the stranger's willingness not only to maintain but to constantly increase the supply of money through the rolling over of existing loans as well as the issuance of new loans. It also creates an imperative for economic growth. At first, the villagers spend their initial rounds paying interest on their debt, but at some point new money has to be lent into the economy in order to service the payment of interest. Since the stranger is only willing to issue new loans where there is a prospect of getting them repaid with interest, the creation of new money requires either the collateralization of existing capital or investment with the prospect of creating new commodities or capital. In other words, the village has to perpetually expand its economy in order to facilitate the perpetual creation of money out of interest-bearing debt.

Debt Drive

Now of course the parable of the eleventh round is an extremely simplified model of the way that economies and banks function. In order to judge the extent to which the mechanisms illustrated in the parable apply to our actual contemporary economy, it is necessary to consider some of the assumptions built into the parable. Our model based on the parable should be regarded as an ideal type in the way defined by Weber: '[The ideal type] is a conceptual construct (*Gedankenbild*) which is neither historical reality nor even the "true" reality.... It has the significance of a purely ideal limiting concept with which the real situation or action is compared and surveyed for the explication of certain of its significant components.'[31]

In the following, we shall be reviewing a number of crucial assumptions made in our model of the parable of the eleventh round. The first assumption is that there is only one kind of bank in the economy. In the actual economy, money is created in the interrelation between central banks and private commercial banks. The stranger in the parable is, however, the only bank in the village economy. The stranger's bank functions as a kind of hybrid between a central bank and a private commercial bank. One the one hand, the stranger has the monopoly on the issuance of the only kind of physical money that circulates in the village. In this sense, he acts as a kind of central bank for the village. On the other hand, the money issued by the stranger is not spent but lent into circulation. The rounds are tokens of deposit money and the villagers have initially borrowed the money from the bank. In this sense, the stranger acts as a private commercial bank.

Rather than discarding the model of the parable of the eleventh round on the premise that it fails to distinguish between central banking and commercial banking, we might argue that this seeming weakness of the model is perhaps one of its strengths. Even if central banks and commercial banks are *de jure* separated, the way that they *de facto* function in our actual contemporary economy means that they work increasingly *as if* they were one bank. This is part of the argument that today we live in the age of post-credit money.[32] The role of most central banks today is not to create new fiat money that is spent directly in the economy by the government. Rather than sovereign creators of money, central banks see themselves as facilitators of the creation of money in private commercial banks. This is covered by the euphemism 'secure financial stability', a variety of which can be found in the mission statement of most central banks. Even when governments do spend money in the economy, this money is typically raised through the selling of bonds, that is, the creation of debt, rather than through the direct printing of money. And even when central banks do create new money, this money is typically not spent directly in the economy but rather lent out to private banks in order to boost their balance sheets and enable them to issue new loans to private agents or to buy government bonds. This means that, even if our ideal typical model of the parable of the eleventh round is indeed a caricature of the actual economy, the assumption that the supply of money can only be increased by the creation of new interest-bearing debt arguably captures, as Weber puts it, a 'significant component' of our contemporary economy.

A second important assumption in our model of the parable of the eleventh round is that the village economy is a closed economy. The scope of production and trade coincides with the scope of the currency zone of the cowhide rounds. In other words, the model does not include imports and exports and there is no economic interaction with other money systems. Again, we might argue that this is a crude simplification in comparison with actual national economies, where there is typically a significant amount of trade with other countries. But, at the same time, the model of the parable of the eleventh round has the advantage of allowing us to observe the whole of the economy at the same time. When we look at accounts of individual national economies, it is not always apparent how trends and events in each economy depend on trends and events in other economies. Of course, it is obvious that the net export of one country implies that other countries are net importers. It is, however, less obvious how the growth and build-up of capital in one country may be dependent on stagnation and indebtedness in other countries. Macroeconomic models that take the perspective of individual countries may function to veil some of the zero-sum game mechanisms in the economy. In contemporary politics, economic growth seems to be posed as a universal solution to all sorts of societal problems. This fantasy is precisely supported by mainstream economic models that only take the perspective of the individual agent, the individual corporation or the economy of an individual nation. In an open economy, the build-up of debt caused by the creation of new money may be exported to other countries, thus evading the accounts of the individual country.

Norway provides an illustrative example of this. Exploiting its vast reserves of oil, the country has grown its economy to the point where it is a huge net creditor. But, for the global economy as a whole, it is structurally impossible for everyone to grow their economy out of debt at the same time. Within one currency zone, economic growth may warrant an expansion of the money supply but this happens only through the creation of an equivalent amount of interest-bearing debt. And between different nations debts and credits ultimately balance out. This means that for one country to export itself out of debt, another country would have to be a net importer, thus running an equivalent deficit.

The gist of our model of the parable of the eleventh round is that when money is created out of interest-bearing debt, it is inherently impossible for the economy to grow itself out of debt. Individual agents may experience wealth and become free of debt and the economy as a whole may even experience an increase in productivity and the build-up of capital. But the money system necessarily increases the amount of debt at a faster rate than the supply of money available to repay the debt. As long as the economy is growing in terms of productive output and build-up of capital, the increasing level of debt may not be a problem. As long as the economy expands, banks are more than willing to keep lending new money into the economy, which is then available to make the current payments of interest and principals on outstanding loans. Such organization of the economy, however, compels it to keep growing if the money system is not to collapse. This Faustian feature of the economy is not immediately obvious when we only look at individual national economies.

According to Baudrillard, the relation between growth and debt is subject to the law of equivalent exchange as long as we focus only on individual agents in the economy. But once we look at the economy as a whole, the relation between growth and debt comes up against the law of impossible exchange. Growth cannot be used to pay off the debt. Banks do not accept chickens in payment of debt and interest. The particular quality of our model of the parable of the eleventh round is exactly that it provides us with such a view of the whole of the economy, which in turn allows us to see how the law of impossible exchange kicks in. While economic growth does indeed result in the production of even more goods and services, these products are only valorized when they are exchanged for money in the market. As long as new money can only be introduced into the economy together with the creation of interest-bearing debt, there will always be a deficit of money, regardless of the amount of goods and services produced in the underlying economy. This is the gist of the parable of the eleventh round. Even if all families were able to double their output of chickens and other commodities, there would not be enough rounds to pay off their debts to the stranger. If we accept the parable as a model of the actual economy, it says that individual countries such as Norway and Germany may be able to grow themselves out of debt but all countries cannot do this at the same time.

Readers with an ear for anti-Semitism will have already noted the third and perhaps most critical assumption in our model of the imaginary village. The parable of the eleventh round is the story of a solidary community, where people live in harmony and mutual collaboration, until it is one day upset by the intrusion of a stranger who wears a hat and lends out money for interest. It is difficult not to notice how the parable resembles the stereotypical narrative of the Jew as the alien element of greed obstructing the balance of an otherwise harmonious community, a narrative which has of course first and foremost been associated with German National Socialism. This resemblance should, however, not prevent us from examining the economic implications of the intrusion of the stranger. The crucial issue is of course not the race nor the beliefs of the stranger but rather his status as someone who is on the margins of the village economy.

One of the reasons why the introduction of the rounds into the village is destined to undermine its economy, or at least force it onto a path of perpetual growth is, that the stranger himself is not a participant in the production, trade and consumption of commodities within the economy. Suppose that, rather than just visiting once every year, the stranger would move into the village after his second visit. And suppose that the stranger, who is now no longer effectively a stranger, consumes chickens, eggs, ham, beer and other commodities produced in the village at the equivalent of 10 rounds annually. In this case, there would not necessarily be a shortage of rounds in circulation to service the debt incurred by the villagers. Rather than being pulled out of circulation, the 10 rounds paid in interest every year would be recycled in the economy as the stranger spends them buying produce in the market. The village would indeed have to make a one-time expansion of its production capacity to provide the surplus demand for commodities brought about by the inclusion of the stranger in the economy but there would be no need for the kind of perpetual growth seen in our initial model.

With this variation of the parable in mind, we might ask if our empirical contemporary economy is closer to the initial ideal type of the parable of the eleventh round, where the stranger is truly a stranger who is separate from the productive economy of the village, or to the alternative ideal type where the stranger is an inhabitant of the village fully integrated into the productive economy. In so far as the stranger represents the function of banking, the question boils down to the relation between banks and the rest of the economy. The issue revolves around the infamous distinction between the financial economy and the productive economy, between Wall Street and Main Street. This is of course a complicated question with no simple and unambitious answer.

It is of course true that banks are integrated with the rest of the economy. There are several ways in which interest and other profits are channelled back into the non-financial economy. For instance, banks pay out dividends to shareholders, as well as salaries and bonuses to employees. This money is in principle available to spend in the non-financial economy, where shareholders, clerks and bank CEOs may use it to buy chickens, cars and other commodities. In addition, banks are also taxed on their profits, which is another way of redistributing money in the non-financial economy. As money is spent back into the non-financial economy, it becomes available for the repayment of debt and interest without the creation of new debt. At the same time, we may also identify a number of possibilities for decoupling the financial economy from the productive or non-financial economy. Rather than paying out money to shareholders, banks may choose to retain the money that is paid back to the bank by debtors and use this for the extension of even more loans and the creation of even more money. Banks may also choose to invest the money in financial markets, where some of it may indeed eventually 'trickle down' to the non-financial economy but substantial sums may just perpetually circulate in a speculative economy, where they remain inaccessible to non-financial agents in the economy. The exponential growth in derivatives trading is sometimes put forward as an indication of such decoupling of the financial from the non-financial economy. As increasing amounts of money are channelled into financial markets, we see that this produces price inflation on securities traded in these markets. This in turn provides the owners of these assets with additional collateral to use for further borrowing and thus further creation of money and debt. It is also the case that some banks aid wealthy individuals and corporations in channelling money into offshore accounts, where it becomes unavailable for taxation.

While it is beyond the scope of this book to provide any extensive review of the trends of financialization, I shall at least venture the hypothesis that our contemporary economy is marked by a trend towards a state of financial apartheid. The popular conception of this apartheid is the distinction between the 99 per cent and the 1 per cent, which was put forward by the Occupy Wall Street Movement. This hypothesis suggests that there is an accumulation and perhaps even congestion of money at one end of the economy, while there is a scarcity of money to service the increasing levels of debt that build up at the other end. If this hypothesis is true, it means that our current economy is closer to the first ideal type, where the stranger is indeed a stranger who is economically decoupled from the productive economy of the village, than to the latter ideal type, where the stranger is an inhabitant of the village. As an increasing decoupling of banking and financial industries from the non-financial sectors of the economy leads to the tensions illustrated in our model of the parable of the eleventh round, the only way of sustaining the system is to continue the injection of ever more money, created out of debt, into the economy. And for this money creation to keep going, we need at least the prospect that our economy can perpetually grow into the future.

The initial purpose of this chapter was to show how the imperative of growth that marks our contemporary economy is not only propelled by needs and desires but also by a pure drive for growth. The drive for growth does not operate on the level of the individual subject but rather on the level of the economy as a whole. The parable of the eleventh round allows us to observe this shift between the two levels. Let us recall the distinction between desire and drive within capitalism, which was quoted at length at the beginning of the chapter:

> At the immediate level of addressing individuals, capitalism of course interpellates them as consumers, as subjects of desire, soliciting in them ever new perverse and excessive desires. /.../ The drive inheres in capitalism at a more fundamental, systemic, level: the drive is that which propels forward the entire capitalist machinery, it is the impersonal compulsion to engage in the endless circular movement of expanded self-reproduction.

When we look at the parable of the eleventh round from the perspective of the individual people in the parable, we can explain economic growth as merely the aggregate function of the desire of individual subjects. The Ericsens have a desire to accumulate capital, try out new solutions in their production and generally optimize and increase their production. This desire is in turn perhaps connected to a desire for status and power in the village or a desire for other commodities that they are able to purchase with their new wealth. We can also empathize with the Sorensens and the Jensens. They have a desire to earn more rounds in order to pay off their debts to the stranger, as well as just generally to get by. Perhaps the Jensens also have a desire to earn enough rounds to help their son buy his own house. The same applies when we look at the actual economy. It is generally possible to explain economic growth by reducing it to the subjective desires of individuals in the economy.

In the parable there is, however, one exception. This is the stranger. What is the desire of the stranger? In so far as he does not appear as a consumer or even as an owner of capital in the village economy, he does not seem to gain anything from his intrusion into the village economy. It is true that ultimately he will be in a position to dominate the village by making demands on all the people who owe him money and all the people who depend on his willingness to extend credit. But even this kind of power seems meaningless as long as he is not part of the village economy himself. Now of course we should be careful how far to extend the logic of the parable. But perhaps the stranger represents the pursuit of money for its own sake – not the desire for the accumulation of capital, nor the desire for commodities to be purchased with money, nor even the desire to become free of debt. There is something enigmatic about the pursuit of money for its own sake. Noam Yuran has provided an elaborate account of the way that conventional economic thinking fails to grasp pure greed because it insists on translating any pursuit of profit into the concept of utility. Yet money pursued for its own sake does not have any utility. Here is how Yuran summarizes his argument:

> At this stage we can finally return to the strange inability of economics to conceptualize greed. The unthinkable nature of greed does not, obviously, result from the fact that greed is a non-economic topic, but rather that it reflects its location beyond or at the limit of the subject. For orthodox economics, greed is unthinkable because it transcends the horizon of utility-seeking individuals. It cannot be fully incorporated within the perspective of individuals. However, for Marx this is precisely what makes it an *objective* economic reality. In this sense the economic oversight of greed is not simply a theoretical mistake. It is an ideological error in the sense of an error that partakes in the social reality it observes. The unthinkable nature of greed is actually part of its structure.[33]

This adds another dimension to the character of the stranger. Not only is he on the margins of the village economy but even the very constitution of his subjectivity is barely comprehensible from the perspective of the villagers. He is not only a stranger but quite simply strange because he incarnates the unthinkable phenomenon of pure greed.

The issue of money creation and the domination of people through the imposition of debt is a recurring theme in many so-called conspiracy theories that try to explain trends and events in the world economy as part of a master plan executed by a small elite of powerful people. The banking system is one of the means through which these elites exercise their powers. In some accounts the conspiracy consists of Jews, in some accounts it consists of aliens from other planets and in some accounts the conspiracy consists of some combination or collaboration between Jews and aliens. Perhaps we can understand these theories as symptoms of this 'unthinkable nature of greed'. In order to explain the existence of pure greed, it is necessary to move beyond the domain of the ordinary subject. So the Jew, the extra-terrestrial or a combination of the two is posed as this alien element which is beyond the horizon of the ordinary human subject. Žižek himself captures this in his discussion of anti-Semitism:

> We must confront ourselves with how the ideological figure of the 'Jew' is invested with our unconscious desire, with how we have constructed this figure to escape a certain deadlock of our desire. /.../ [T]he anti-Semitic idea of Jew has nothing to do with Jews; the ideological figure of a Jew is a way to stitch up the inconsistency of our own ideological system.[34]

There are of course plenty of reasons to be highly sceptical of theories that resort to the existence of a conspiracy of Jews or aliens as a way of explaining the world. But we may still concur with the underlying philosophical point that, in order to understand money and the banking system, we need to include some extra-subjective element in our explanation. Fortunately, Žižek provides us with concepts that allow us to include this element in our explanation without resorting to blaming Jews or aliens. This brings us back to the concept of drive. The stranger in the parable functions to incarnate pure drive. As we have seen, the intrusion of the stranger into the village economy seems to put it on a path of forced perpetual growth that may lead to the community's ultimate destruction. Žižek captures this feature of drive by defining it as 'death drive':

> [D]rive as such is death drive: it stands for an unconditional impetus which disregards the proper needs of the living body and simply

battens on it. It is as if some part of the body, an organ, is sublimated, torn out of its bodily context, elevated to the dignity of the Thing and thus caught in an infinitely repetitive cycle, endlessly circulating around the void of its structuring impossibility.[35]

A transposition of this definition of drive onto the domain of the economy might look like this:

The drive of contemporary capitalism is debt drive: it stands for an unconditional impetus which disregards the proper needs of the living society and simply battens on it. It is as if some part of the economy, the bank, is sublimated, torn out of its societal context, elevated to the dignity of the Thing and thus caught in an infinitely repetitive cycle, endlessly circulating around the void of the impossibility of debt redemption.

Conclusion: 'It's the Money, Stupid!'

The philosophers have only interpreted the world, in various ways; the point is to change it.[1]

The threat today is not passivity, but pseudo-activity, the urge to 'be active', to 'participate', to mask the nothingness of what goes on. People intervene all the time, 'do something'; academics participate in meaningless debates, and so on. The truly difficult thing is to step back, to withdraw.[2]

It would be nice to conclude this book with a series of practical answers to the challenges facing the world in our current times of economic and ecological crisis. However, such recommendations would undermine not only the structure but the whole idea of the book. We started out with the Lenin joke that was rewritten into four different responses to the choice between economy and ecology. As each of the three parts of this book corresponds to one of these responses, it would make sense to conclude with the fourth Leninist option. But, as already indicated in the introductory chapter, the analytical engagement with the concepts of ecology and economy constitutes, in itself, this fourth option. In psychoanalysis, the task of the analyst is not to come up with practical solutions to the perceived problems of the analysand. The idea is rather for the analyst to engage the analysand in a common analysis of the issues that prevent the analysand from realizing his or her own solutions to his or her own problems. The ambition of eco-analysis, which has been unfolded in this book, is to do the same thing with regard to 'the eco'. Of course, this ambition is less straightforward since the meaning of the notion of the eco is anything but self-evident.

We have already seen how Žižek defines the task of philosophy not as providing answers but rather as showing how our perception of a problem is sometimes part of the problem itself. This is also the gist of the above quote on passivity and pseudo-activity. If we look at the field of ecological politics today, Žižek's critique does appear relevant. This field of politics is marked by a pervasive sense of urgency generated by calls to 'Act now!' and 'Save the planet!' At first glance, these politics seem to have fully adopted Marx's eleventh thesis on Feuerbach which is also quoted above. The point now is to change the world. The premise of Marx's thesis is, however, that the philosophers have *already* interpreted the world. This raises the question of whether contemporary ecological politics is informed by a sufficient analysis of the situation. There is of course an abundance of scientific analyses documenting trends of global climate change, biodiversity decrease, natural resource depletion and so on. But as argued throughout this book, such a view of the world fails to take into account the peculiar ontology of the human subject. In this sense, ecology is a contemporary version of materialism. Now the point in Marx's critique on Feuerbach was that, while being a materialist, Feuerbach did not live up to Marx's definition of materialism. Using Žižek, we may summarize our critique of ecology in a similar fashion. 'Materialism', says Žižek:

> is not the direct assertion of my inclusion in objective reality (such an assertion presupposes that my position of enunciation is that of an external observer who can grasp the whole of reality); rather, it resides in the reflexive twist by means of which I myself am included in the picture constituted by me – it is this reflexive short circuit, this necessary redoubling of myself as standing both outside and inside my picture, that bears witness to my 'material existence'. Materialism means that the reality I see is never 'whole' – not because a large part of it eludes me, but because it contains a stain, a blind spot, which signals my inclusion in it.[3]

The Apollo 17 photograph provides exactly such an external view of the 'whole of reality'. The problem is of course that it fails to make the reflexive twist by means of which the living subject, itself an observer of reality, is included in the picture. This failure to include the human subject in the account of reality is one of the reasons why the sense of urgency and the incessant calls for immediate action within the field of ecological politics largely fail to mobilize truly transformative forms of collective action. That is, of course, not to say that the politics of ecology is not already having an effect. But these effects may not always be as unambiguously positive as we would like to think. If we do not properly understand its underlying notions about the relations between Man and Nature and its implicit conception of the ontology of the human subject, ecology may develop into just another form of totalitarian ideology that serves to maintain or even strengthen existing forms of domination and exploitation. Again, here is how Žižek elaborates on the point:

> If there is a lesson to be learned from the so-called 'totalitarian' experience, it is that the temptation is exactly the opposite: the danger of imposing, in the absence of any divine limit, a new pseudo-limit, a fake transcendence on behalf of which I act (from Stalinism to religious fundamentalism). Even ecology functions as ideology the moment it is evoked as a new Limit: it has every chance of developing into the predominant form of the ideology of global capitalism, a new opium for the masses replacing the declining religion, taking over the old latter's fundamental function, that of assuming an unquestionable authority which can impose limits. The lesson this ecology constantly hammers into us is our finitude: we are not Cartesian subjects extracted from reality, we are finite beings embedded in a bio-sphere which vastly exceeds our own horizons.[4]

There is a balance to be struck. On the one hand, we should of course take into account the way that contemporary forms of production and consumption have destructive effects on the natural environment which constitutes our habitat. On the other hand, this should not lead us to the reduction of the human subject to merely just another piece of nature. Metaphysics is an inherent part of the nature of the human being. The task of eco-analysis is to navigate this split, which is constitutive of the eco. The challenge is to make ecology into a viable ideology that informs the way society is constituted, while at the same time curbing its potential to develop into a new totalitarianism.

If ecology is an, albeit insufficient, exponent of materialist thinking, then mainstream economics is perhaps one of the most forceful exponents of idealism today. We have seen how neo-classical economics is based on a market theory of value. This means that price and value conflate and neo-classical economic theory is able to speak about value only in so far as it is priced in monetary terms in the market. The object of economic analysis is the market's perception of the economy, while it is impossible to speak of the economy in itself as being anything beyond this perception. Within this paradigm of economic thinking, the concept of growth also relies on an idealist epistemology. Economic growth stands for the increase in productive output that is priced in the market and exchanged for money. The theory is thus unable to distinguish between growth through increase in productivity and growth through the incorporation of production, labour and capital that was already in existence in a more general non-monetary economy.

As soon as we have realized the inherent idealism of contemporary economics, it is easy to mobilize a range of critical points for this paradigm of thinking by invoking a materialist epistemology. Such critical points would include precisely the failure of mainstream economics to take into account those domains of the world that are not incorporated into the market economy. This is essentially what the field of ecological economics is all about. There is no reason to discard these highly relevant points of critique, and indeed I have repeated several of them throughout the book. The problem is, however, that if we do not proceed beyond this kind of critique, we remain trapped in the classic dialectics between materialism and idealism. When the materialist ecological economist argues that the idealist neo-classical economist fails to take into account the value of natural capital that is destroyed through economic growth, the neo-classical economist will argue that it is only a matter of time before the market will eventually adapt and pricing mechanisms for 'externalities' will emerge. He might even add that, in order to speed up this process of market adaptation, we need to subject natural capital to private ownership so that it can be traded and priced. The materialist ecological economist may then counter that this is merely going to accelerate the exploitation of whatever intact natural capital is still left. And the idealist neo-classical economist may then resort to the pragmatic argument that, if we want to do something about the problem, there is no way of working around the mechanisms of the market. Only when things have a price do they have a value that is perceptible to the market.

The reason why this dialectic is problematic is because it tends to lead into a kind of unproductive or even false synthesis. This is what we find in the notion of 'green growth' or similar political concepts that purport to overcome the inherent contradictions between ecology and economy. We are of course now within the third option in the structure set out by the Lenin joke. Proponents of such policies of 'green growth' or 'sustainable development' tend to overlook any contradiction between ecology and economy that cannot be overcome through a series of adjustments to our current modes of production as well as our habits of consumption. We can frame this position through another one of Žižek's jokes:

> There is an (apocryphal, for sure) anecdote about the exchange of telegrams between German and Austrian army headquarters in the

middle of the First World War: the Germans sent the message 'Here, on our part of the front, the situation is serious, but not catastrophic', to which the Austrians replied 'Here, the situation is catastrophic, but not serious.' Is this not more and more the way many of us, at least in the developed world, relate to our global predicament? We all know about the impending catastrophe – ecological, social – but we somehow cannot take it seriously.[5]

As we have already explored, policies of 'green growth' recognize the dire prospects that lie ahead for humanity if patterns of growth continue along their current path. At the same time, the proposals for change presented as part of these policies are typically still well within the premises of the existing system of capitalism. Even if there is sometimes talk of a 'green revolution', this type of revolution is a far cry from the type of revolution envisaged by Marx or any other of the great revolutionary thinkers. Al Gore stands out as emblematic of this position. As we have seen, in his famous talk he initially presents us with the 'inconvenient truth' that humanity is heading for a catastrophe caused by global warming but then ends on the comforting note that this catastrophe can be prevented if we shift to using compact fluorescent light bulbs, buy hybrid cars or even just encourage our friends to watch the film. This is in effect saying: 'The situation is catastrophic but not serious.' In Obama's 2013 Climate Change speech, this tragic mechanism is repeated as farce when the increased US domestic extraction of natural gas through fracking technologies is emphasized as part of the transformation of America into a more sustainable economy.

In order to avoid the deadlock of the materialist ecology and idealist-economy dialectic, we should emphasize another type of critique of the economy that does not so much rest on an ecological perspective but rather targets economics on its own premises. This would be a critique that is more in line with the Kantian definition of the word. Throughout our analysis of economics in general and neo-classical economics in particular, we have seen how the object of money itself seems to be absent from key areas of economic thinking. We have seen how classical economics fails to recognize how the introduction of money into the economy is constitutive of the shaping of the economic subject. We have also seen how neo-classical economics fails to recognize money itself as the transcendental schemata enabling certain observations while at the same time making other observations impossible. And we have seen how the imperative of economic growth is not entirely, perhaps not even primarily, driven by a demand for more commodities and services but simply by a demand for ever more money to pay back the debt that is being created every time new money is created. The point of such critique is to demonstrate that not only is the organization of our current mode of production and consumption detrimental to rainforests, polar bears and the sustenance of all kinds of ecosystems, it is even at odds with the premises on which mainstream economics itself is resting. Noam Yuran moves one step further along these lines as he speculates how

> the speed with which the ecological message has spread within affluent societies, where alternative economic regimes are far from sight, raises the suspicion that the ecological discourse is to some extent a replacement for an absent, fundamental social and economic debate. /.../ [B]y presenting the ecological threat as the reason why fundamental change in the economy is required, these critics actually turn ecology into a mask that hides the real and necessary critical discourse about the economy.[6]

There is an implicit anthropology running through the history of economic thinking. This is the idea that the human subject always wants more. In eighteenth-century France, when the physiocrats were laying down the groundwork for modern economics, this assumption would seem reasonable in so far as large parts of the population were living in conditions where they had to struggle to merely put food on the table. And in eighteenth- and nineteenth-century Britain, when the classical economists were writing, it was also not an unrealistic assumption. Even in the age of industrialization, economic growth was still largely a matter of providing basic necessities for people. But as we move forward through the twentieth and into the twenty-first century, general affluence is becoming the norm in many economies. This means that the assumption that perpetual economic growth is still just the expression of 'unlimited needs and wants' (Samuelson) inherent in human nature seems to become gradually more and more absurd. The point here is not to suggest that there is a natural limit to consumption. It is to say that, rather than taking the seemingly unlimited needs and wants of society for granted as an anthropological constant, they should be included as part of the study of the economy. This is exactly one of the crucial differences between economics and eco-analysis.

When the constitution of money is included in the study of the economy, we see how money itself may function as a driver for growth. This is particularly the case within the paradigm of post-credit money. Without discarding entirely the proposition that people want growth, our analysis aims to emphasize the point that today it is perhaps rather money itself that wants growth. This point is important in so far as it opens up new avenues for thinking about solutions to the current crises.

In our reformulation of the Lenin joke, we found three plus one different responses to the dual crisis of ecology and economy. First, there is the left-wing ecologist who would ignore the economic crisis and solve the ecological crisis by limiting consumption. Second, there is the right-wing economist who would ignore or even deny the ecological crisis and ride it out through sustained patterns of consumption. This solution rests on a belief that capitalism is a self-sustaining system that is able to repair itself. And, third, there is the progressive liberal politician who would solve both crises through a transformation of the economy into a green capitalism. Within this solution, we find a steadfast belief in the progress of 'knowledge or technology'.

The fourth solution was of course the 'Leninist' approach of rejecting both ecology and economy in order to go through the process of eco-analysis. What this analysis has hopefully opened up is the perspective on money. The concluding proposal of this book is that, rather than trying to change the habits of consumers, ignoring the ecological crisis and letting capitalism work itself out, or sitting back and hoping that science will solve our problems, we should think through the things that may be achieved by changing our monetary system.

Notes

Introduction: Lenin at the Supermarket

1. Žižek, *Violence*, 7.
2. Žižek, *The Sublime Object of Ideology*, 69, 169.
3. Ibid., 122.
4. Žižek, *The Parallax View*, 26.
5. Žižek, *Looking Awry*, 3–20.

Chapter 1 The Balance of Nature

1. Žižek, *Less than Nothing*, 372–3.
2. Egerton, 'Changing Concepts of the Balance of Nature'.
3. Cited in Egerton, *Roots of Ecology*, 2.
4. Bjerg, *Etik Uden Moral*, 100–18.
5. Cited in Egerton, 'Changing Concepts of the Balance of Nature', 333.
6. Cited in ibid., 336.
7. Smith, *The Theory of Moral Sentiments*, 187.
8. Smith, *The Wealth of Nations*, 484–5.
9. Žižek, *The Ticklish Subject*, 109–10.
10. Darwin, *The Origin of Species*, 50–1.
11. Ibid., 58.
12. Foster, *The Vulnerable Planet*, 50–67.
13. Marsh, *Man and Nature*, 36.
14. Haeckel, *Gesammelte Populäre Vorträge aus dem Gebiete der Entwickelungslehre*.
15. Cooper, *The Science of the Struggle for Existence*.
16. Ricardo, *The Principles of Political Economy and Taxation*, 81.

17 Leibniz, *Theodicy*.
18 Žižek, *The Sublime Object of Ideology*, 123.
19 Ibid., 45.

Chapter 2   Ecology Beyond Biology

1   Forbes, 'The Lake as a Microcosm', 90.
2   Tansley, 'The Use and Abuse of Vegetational Concepts and Terms', 299.
3   Cooper, *The Science of the Struggle for Existence*, 62.
4   Vernadsky, 'Problems of Biogeochemistry II', 484.
5   Ibid.
6   Ibid., 483.
7   Bachelier, *Theory of Speculation*.
8   Fama, 'Efficient Capital Markets', 383.
9   Lovelock and Margulis, 'Atmospheric Homeostasis by and for the Biosphere', 3.
10 United Nations, *Report of the World Commission on Environment and Development: Our Common Future*, 11.
11 Story is retold in Žižek, *The Sublime Object of Ideology*, 58.
12 Meadows et al., *The Limits to Growth – A Report for THE CLUB OF ROME'S Project on the Predicament of Mankind*.
13 United Nations, *Report of the World Commission on Environment and Development: Our Common Future*, 26.
14 Ibid.
15 Žižek, 'Nature and Its Discontents', 58–9.
16 Hopenhagen, 'Welcome to Hopenhagen'.
17 United Nations, *United Nations Rio+20 The Future We Want*, 1.
18 Žižek, *Looking Awry*, 33.
19 Ibid., 45; Žižek, *The Ticklish Subject*, 60.

Chapter 3   How *is* the Economy?

1   International Monetary Fund, *IMF World Economic Outlook (WEO) – Hopes, Realities, and Risks*, 1.1.
2   See Bjerg, *Making Money – The Philosophy of Crisis Capitalism*, 3–4 for a further explication of Heidegger's posing of this question.
3   Heidegger, 'Der Feldweg (1949)', 38.
4   [Meanwhile, the hardness and scent of the oakwood began to speak more distinctly of the slowness and steadiness with which the tree grows. The oak itself said that. 'In such growth alone is grounded that which lasts and fructifies'; growing means: to open oneself to the expanse of the heavens as one takes root in the darkness of the earth; that everything genuine thrives only when man is both in right measure:

ready for the claim of the highest heavens and elevated in the protection of the bearing earth.]
5   Quesnay, *Tableau Économique*.
6   Turgot, *Reflections on the Formation and Distribution of Wealth* Rfl. 15–16.
7   Ibid. Rfl. 36.
8   Ibid. Rfl. 34, 36.
9   Lopdrup-Hjorth, *The Value of Co-Creation*, 77.
10 See also Bjerg, *Making Money – The Philosophy of Crisis Capitalism*, 20–8.
11 Žižek, *The Parallax View*, 26.
12 Quesnay, *Tableau Économique*.
13 Quesnay, *Tableau Économique*.
14 Turgot, *Reflections on the Formation and Distribution of Wealth* Rfl. 120.
15 Smith, *The Wealth of Nations*, 32.
16 Ibid., 33.
17 Ibid., xxiii.
18 Lopdrup-Hjorth, *The Value of Co-Creation*, 80–1.
19 Smith, *The Wealth of Nations*, 731.
20 Mirowski, *More Heat than Light*, 165.
21 Lacan, 'The Mirror Stage as Formative of the Function of the I as Revealed in Psychoanalytic Experience', 848.
22 Ricardo, *The Principles of Political Economy and Taxation*, 14.
23 Ibid., 33.
24 Ibid., 39.
25 Ibid., 38.
26 Smith, *The Wealth of Nations*, 3.
27 Ibid., 7–11.
28 Žižek, *The Plague of Fantasies*, 15.
29 Ibid., 14.
30 Smith, *The Wealth of Nations*, 24.

Chapter 4   The Market Theory of Value

1   Gale, 'The Law of Supply and Demand', 155.
2   Žižek, *The Sublime Object of Ideology*, 45.
3   Baudrillard, *Impossible Exchange*, 3–4.
4   Ibid., 5–6.
5   Friedman, 'The Methodology of Positive Economics', 7, 8, 14.
6   Žižek, *Looking Awry*, 6.
7   Mankiw, Phelps and Romer, 'The Growth of Nations', 308.
8   Solow, 'A Contribution to the Theory of Economic Growth', 65.
9   Swan, 'Economic Growth and Capital Accumulation'.
10 Solow, 'A Contribution to the Theory of Economic Growth', 66.

11 Ibid., 67.
12 Mankiw, Phelps and Romer, 'The Growth of Nations', 638.
13 Ibid., 642.
14 Daly, *Steady-State Economics*.
15 Solow, 'A Contribution to the Theory of Economic Growth', 67.
16 Lacan, 'The Mirror Stage as Formative of the Function of the I as Revealed in Psychoanalytic Experience', 848.
17 Barro and Sala-i-Martin, *Economic Growth*.
18 See for instance Mankiw, *Principles of Microeconomics*, 25.
19 Daly, *Beyond Growth*, 47.
20 Daly, *Beyond Growth*.
21 Harris and Roach, *Environmental and Natural Resource Economics*.

Chapter 5   The Fantasy of Growth without Bounds

1   Barro and Sala-i-Martin, *Economic Growth*, 17.
2   Romer, 'Increasing Returns and Long-Run Growth', 1003.
3   Barro and Sala-i-Martin, *Economic Growth*, 24.
4   Žižek, *Contingency, Hegemony, Universality*, 90.
5   Žižek, *The Plague of Fantasies*, 9.
6   Žižek, *The Fragile Absolute*, 28.
7   Mill, *Principles of Political Economy with Some of Their Applications to Social Philosophy*, bk. IV.6.2.
8   Keynes, 'The Economic Possibilities of Our Grandchildren'.
9   The definition is typically attributed to Paul Samuelson (1970) but the exact origin of the quote could not be retrieved.
10 Barro and Sala-i-Martin, *Economic Growth*, 24.
11 Žižek, *Looking Awry*, 6.
12 Žižek, *In Defense of Lost Causes*, 457.
13 Wilkinson and Pickett, *The Spirit Level*.
14 Žižek, *How to Read Lacan*, 23.
15 Ibid., 24.
16 Babcock, 'The Impact of US Biofuel Policies on Agricultural Price Levels and Volatility'.
17 Baudrillard, *For a Critique of the Political Economy of the Sign*, 207.
18 Mankiw and Taylor, *Macroeconomics* (European edn), 202.
19 Žižek, *The Ticklish Subject*, 101.
20 World Bank, 'GDP per Capita, PPP (current International $) | Data | Table'.
21 Kuznets, 'National Income, 1929–1932', 7.
22 Žižek, *The Sublime Object of Ideology*, 29.
23 Ibid., 45.
24 Barro and Sala-i-Martin, *Economic Growth*, 24.
25 Wittgenstein, *Tractatus Logico-Philosophicus*, para. 7.
26 Mankiw, Phelps and Romer, 'The Growth of Nations', 275.

Chapter 6 The Need to Grow

1. The Danish Government, *Responsible Growth: Action Plan for Corporate Social Responsibility 2012–2015*, 3–4.
2. Obama, *Remarks by the President on Climate Change*.
3. OECD, *Towards Green Growth*, 18.
4. Brand, 'Green Economy and Green Capitalism'.
5. Boltanski and Chiapello, *The New Spirit of Capitalism*.
6. Sachs, *Planet Dialectics*; Tanuro, *Green Capitalism*; Brand, 'Green Economy – the Next Oxymoron?'
7. Pierre-Louis, *Green Washed*.
8. Daly, *Steady-State Economics*; Jackson, *Prosperity without Growth*; Czech, *Supply Shock*.
9. Rockström et al., 'Planetary Boundaries'.
10. Žižek, *Year of Distraction*.
11. OECD, *Towards Green Growth*, 9.
12. Rowbotham, *The Grip of Death*; Binswanger, *Die Wachstumsspirale*.
13. Molina et al., *What We Know*, 1–4.
14. Sloterdijk, *Spheres. Vol. 1*, 83.
15. Molina et al., *What We Know*, 1.
16. United Nations, *Report of the World Commission on Environment and Development: Our Common Future*, 37.
17. Monsanto, 'Monsanto | Our Commitment to Sustainable Agriculture'.
18. Robin, *The World According to Monsanto*.
19. Lacan, 'The Subversion of the Subject and the Dialectic of Desire in the Freudian Unconscious', 689.
20. Godfray et al., 'Food Security'.
21. Lacan, 'The Function and Field of Speech and Language in Psychoanalysis', 312.
22. Žižek, *Looking Awry*, 31–2.
23. Žižek, *The Fragile Absolute*, 149, 150.

Chapter 7 The Desire to Grow

1. Žižek, *The Plague of Fantasies*, 29.
2. Smith, *The Wealth of Nations*, 24–5.
3. Bjerg, *Making Money – The Philosophy of Crisis Capitalism*, 90–6.
4. Graeber, *Debt*, 28.
5. See also Humphrey, 'Barter and Economic Disintegration'.
6. Eisenstein, *Sacred Economics*, 3–18.
7. Žižek, *The Fragile Absolute*, 28.
8. Žižek, *The Parallax View*, 17–18.
9. Žižek, *The Plague of Fantasies*, 39.
10. Obama, *Remarks by the President on Climate Change*.
11. European Commission, *LIFE Creating Jobs and Skills*, 3.

12 Mill, *Principles of Political Economy with Some of Their Applications to Social Philosophy*, 7–8.
13 Mishkin, *The Economics of Money, Banking, and Financial Markets*, 45.
14 Boltanski and Chiapello, *The New Spirit of Capitalism*.
15 Bjerre, 'Miljøbevægelsens Nye Ånd'.
16 Fraser, *The Fortunes of Feminism*, 222.
17 Danmarks Statistik, *Kvinder & Mænd 2011*.
18 Common and Stagl, *Ecological Economics*, 1.
19 Berry, *What Matters?*, 32.
20 Fox and Fimeche, 'Global Food'.
21 Berry, *What Matters?*, 77.
22 Lacan, *Television*, 38.
23 Cleveland, *Biophysical Economics*.
24 Žižek, *The Plague of Fantasies*, 29.
25 Fraser, *The Fortunes of Feminism*, 225–6.
26 Mauss, *The Gift*, 4.
27 Žižek, *How to Read Lacan*, 11–12.

Chapter 8 The Drive for Growth

1 Žižek, *Less than Nothing*, 496–7.
2 Žižek, *The Fragile Absolute*, 28.
3 Žižek, *The Plague of Fantasies*, 30.
4 Bjerg, *For Tæt På Kapitalismen: Ludomani, Narkomani Og Købemani*.
5 Bjerg, 'Too Close to the Money'.
6 Bjerg, *Poker – The Parody of Capitalism*, 150–3.
7 Quoted in Alvarez, *The Biggest Game in Town*, 114.
8 United Nations Environment Programme, *Decoupling Natural Resource Use and Environmental Impacts from Economic Growth*, xi.
9 OECD, *Sustainable Development: Indicators to Measure Decoupling of Environmental Pressure from Economic Growth*.
10 Žižek, *The Puppet and the Dwarf*, 96.
11 Daly, *Steady-State Economics*; Jackson, *Prosperity without Growth*; Czech, *Supply Shock*.
12 Dittrich et al., *Green Economies around the World?*
13 Georgescu-Roegen, *The Entropy Law and the Economic Process*.
14 Cleveland, *Biophysical Economics*.
15 Georgescu-Roegen, 'Myth about Energy and Matter'.
16 Bjerg, *Making Money – The Philosophy of Crisis Capitalism*, 96–100.
17 Ibid., 155–67.
18 Ibid., 167–83.
19 Ryan-Collins et al., *Where Does Money Come From? A Guide to the UK Monetary and Banking System*; Jackson and Dyson, *Modernising Money*.

20 Werner, *New Paradigm in Macroeconomics*, 161–80.
21 Davies and Green, *Banking on the Future*, 26–7.
22 Bjerg, *Making Money – The Philosophy of Crisis Capitalism*, 252–3.
23 Ibid., 139.
24 Ibid., 22.
25 Ibid., 124–31.
26 Lietaer, *The Future of Money*, 50–3.
27 Eisenstein, *Sacred Economics*, 95–7.
28 Žižek, *The Sublime Object of Ideology*, 3.
29 Žižek, *The Parallax View*, 18.
30 Žižek, *Less than Nothing*, 132.
31 Weber, '"Objectivity" in Social Science and Social Policy', 93.
32 Bjerg, *Making Money – The Philosophy of Crisis Capitalism*, 155–250.
33 Yuran, *What Money Wants*, 31.
34 Žižek, *The Sublime Object of Ideology*, 48.
35 Žižek, *The Plague of Fantasies*, 31.

Conclusion: 'It's the Money, Stupid!'

1 Marx, *Selected Writings*, 101.
2 Žižek, *Violence*, 183.
3 Žižek, *The Parallax View*, 17.
4 Žižek, *Less than Nothing*, 979.
5 Ibid., 996–7.
6 Yuran, *What Money Wants*, 175–6.

Bibliography

Alvarez, Al. 1983. *The Biggest Game in Town*. London: Bloomsbury.
Babcock, Bruce A. 2012. 'The Impact of US Biofuel Policies on Agricultural Price Levels and Volatility'. *China Agricultural Economic Review* 4(4): 407–26.
Bachelier, Louis. 1900. *Louis Bachelier's Theory of Speculation: The Origins of Modern Finance*. Princeton: Princeton University Press.
Barro, Robert J. and Sala-i-Martin, Xavier I. 2003. *Economic Growth*. 2nd edn. Massachusetts: MIT Press.
Baudrillard, Jean. 1981. *For a Critique of the Political Economy of the Sign*. St Louis: Telos Press Publishing.
Baudrillard, Jean. 2001. *Impossible Exchange*. London: Verso.
Berry, Wendell. 2010. *What Matters?* Berkeley, CA: COUNTERPOINT.
Binswanger, Hans Christoph. 2013. *Die Wachstumsspirale: Geld, Energie und Imagination in der Dynamik des Marktprozesses*. Marburg: Metropolis-Verlag.
Bjerg, Ole. 2008. *For Tæt På Kapitalismen: Ludomani, Narkomani Og Købemani*. København: Museum Tusculanum Press.
Bjerg, Ole. 2009. 'Too Close to the Money'. *Theory, Culture & Society* 26(4): 47–66.
Bjerg, Ole. 2010. *Etik Uden Moral*. København: Museum Tusculanum Press.
Bjerg, Ole. 2011. *Poker – The Parody of Capitalism*. Ann Arbor: University of Michigan Press.
Bjerg, Ole. 2014. *Making Money – The Philosophy of Crisis Capitalism*. London: Verso.
Bjerre, Henrik Jøker. 2009. 'Miljøbevægelsens Nye Ånd'. *Dansk Sociologi* 19(2): 127–47.
Boltanski, Luc and Chiapello, Ève. 2005. *The New Spirit of Capitalism*. London: Verso.

Brand, Ulrich. 2012. 'Green Economy and Green Capitalism: Some Theoretical Considerations'. *Journal für Entwicklungspolitik* 28(3): 118–37.
Brand, Ulrich. 2012. 'Green Economy – the Next Oxymoron? No Lessons Learned from Failures of Implementing Sustainable Development'. *GAIA –Ecological Perspectives for Science and Society* 21(1): 28–32.
Cleveland, Cutler J. 1999. *Biophysical Economics: From Physiocracy to Ecological Economics and Industrial Ecology*. Cheltenham, UK: Edward Elgar Publishing.
Common, Michael and Stagl, Sigrid. 2005. *Ecological Economics: An Introduction*. Cambridge: Cambridge University Press.
Cooper, Gregory J. 2007. *The Science of the Struggle for Existence: On the Foundations of Ecology*. Cambridge: Cambridge University Press.
Czech, Brian. 2013. *Supply Shock: Economic Growth at the Crossroads and the Steady State Solution*. Gabriola, BC: New Society Publishers.
Daly, Herman E. 1991. *Steady-State Economics*. Washington, DC: Island Press.
Daly, Herman. 1996. *Beyond Growth: The Economics of Sustainable Development*. Boston: Beacon Press.
Danish Government. 2012. *Responsible Growth: Action Plan for Corporate Social Responsibility 2012–2015*. Copenhagen: Erhvervsstyrelsen.
Danmarks Statistik. 2011. *Kvinder & Mænd 2011*. TemaPubl. København: Danmarks Statistik.
Darwin, Charles. 1859. *The Origin of Species*. Hertfordshire, UK: Wordsworth Editions.
Davies, Howard, and Green, David. 2010. *Banking on the Future: The Fall and Rise of Central Banking*. Princeton: Princeton University Press.
Dittrich, Monika, Giljum, Stefan, Lutter, Stephan and Polzin, Christine. 2012. *Green Economies around the World? Implications of Resource Use for Development and the Environment*. Vienna: SERI-Sustainable Europe Research Institute.
Egerton, Frank N. 1973. 'Changing Concepts of the Balance of Nature'. *Quarterly Review of Biology* 48(2): 322–50.
Egerton, Frank N. 2012. *Roots of Ecology: Antiquity to Haeckel*. University of California Press.
Eisenstein, Charles. 2011. *Sacred Economics: Money, Gift, & Society in the Age of Transition*. Berkeley: Evolver Editions.
European Commission. 2013. *LIFE Creating Jobs and Skills*. Luxembourg: Publications Office of the European Union.
Fama, Eugene F. 1970. 'Efficient Capital Markets: A Review of Theory and Empirical Work'. *Journal of Finance* 25(2): 383–417.
Forbes, Stephen A. 1887. 'The Lake as a Microcosm'. *Bulletin of the Scientific Association*, 77–87.
Foster, John Bellamy. 1999. *The Vulnerable Planet: A Short Economic History of the Environment*. New York: Monthly Review Press,
Fox, Tim, and Fimeche, C. 2013. 'Global Food: Waste Not, Want Not'. *Institute of Mechanical Engineers* (Jan.).

Fraser, Nancy. 2013. *The Fortunes of Feminism: From Women's Liberation to Identity Politics to Anti-Capitalism*. London: Verso Books.
Friedman, Milton. 1953. 'The Methodology of Positive Economics', in Daniel M. Hausman (ed.), *The Philosophy of Economics: An Anthology*, 2: Cambridge: Cambridge University Press, pp. 180–213.
Gale, David. 1955. 'The Law of Supply and Demand'. *Mathematica Scandinavica* 3: 155–69.
Georgescu-Roegen, Nicholas. 1971. *The Entropy Law and the Economic Process*. Cambridge, MA: Harvard University Press.
Georgescu-Roegen, Nicholas. 1979. 'Myth about Energy and Matter'. *Growth and Change* 10(1): 16–23.
Godfray, H. Charles J., Beddington, John R., Crute, Ian R. et al. 2010. 'Food Security: The Challenge of Feeding 9 Billion People'. *Science* 327(5967): 812–18.
Graeber, David. 2011. *Debt: The First 5,000 Years*. New York: Melville House.
Haeckel, Ernst Heinrich Philipp August. 1866. *Gesammelte Populäre Vorträge aus dem Gebiete der Entwickelungslehre*. Bonn: Emil Strauss.
Harris, Jonathan M. and Roach, Brian. 2014. *Environmental and Natural Resource Economics: A Contemporary Approach*. Armonk, NY; Routledge.
Heidegger, Martin. 1949. 'Der Feldweg (1949)'. In *Denkerfahrungen 1910–1976*. Frankfurt am Main: Vittorio Klostermann, pp. 37–40.
Hopenhagen. 'Welcome to Hopenhagen', available at www.hopenhagen.org.
Humphrey, Caroline. 1985. 'Barter and Economic Disintegration'. *Man* 20(1): 48–72.
International Monetary Fund. 2013. *IMF World Economic Outlook (WEO): Hopes, Realities, and Risks*. International Monetary Fund.
Jackson, Andrew and Dyson, Ben. 2013. *Modernising Money: Why Our Monetary System Is Broken and How It Can Be Fixed*. London: Positive-Money.
Jackson, Tim. 2009. *Prosperity without Growth: Economics for a Finite Planet*. London: Routledge.
Keynes, John Maynard. 1930. 'The Economic Possibilities of Our Grandchildren'. In *Collected Writings of J. M. Keynes*, Vol. IX. London: Macmillan for the Royal Economic Society.
Kuznets, Simon. 1934. 'National Income, 1929–1932'. In *National Income, 1929–1932*. NBER, pp. 1–12.
Lacan, Jacques. 1949. 'The Mirror Stage as Formative of the Function of the I as Revealed in Psychoanalytic Experience'. In *Écrits*. London: W. W. Norton.
Lacan, Jacques. 1960. 'The Subversion of the Subject and the Dialectic of Desire in the Freudian Unconscious'. In *Ecrits*. London: W. W. Norton.
Lacan, Jacques. 1966. 'The Function and Field of Speech and Language in Psychoanalysis'. In *Ecrits*. London: W. W. Norton.
Lacan, Jacques. 1990. *Television: A Challenge to the Psychoanalytic Establishment*. London: W. W. Norton.

Leibniz, Gottfried Wilhelm. 1710. *Theodicy*. New York: Cosimo Books.
Lietaer, Bernard. 2002. *The Future of Money: Creating New Wealth, Work and a Wiser World*. London: Random House Business.
Lopdrup-Hjorth, Thomas. 2013. *The Value of Co-Creation*. Frederiksberg: Copenhagen Business School.
Lovelock, James E. and Margulis, Lynn. 1974. 'Atmospheric Homeostasis by and for the Biosphere: The Gaia Hypothesis'. *Tellus* 26(1–2): 2–10.
Mankiw, N. Gregory. 2011. *Principles of Microeconomics*. 6th edn. Mason, OH: Cengage Learning.
Mankiw, N. Gregory, Phelps, Edmund S. and Romer, Paul M. 1995. 'The Growth of Nations'. *Brookings Papers on Economic Activity* 1995(1): 275–326.
Mankiw, N. Gregory and Taylor, Mark P. 2008. *Macroeconomics*, European edn. New York: Worth Publishers.
Marsh, George P. 1864. *Man and Nature; or, Physical Geography as Modified by Human Action*. New York: C. Scribner.
Marx, Karl. *Selected Writings*. 1845. Cambridge: Hackett Publishing.
Mauss, Marcel. 1950. *The Gift: Forms and Functions of Exchange in Archaic Societies*. London: Routledge.
Meadows, Donella H., Meadows, Dennis L., Randers, Jørgen and Behrens III., William W. 1972. *The Limits to Growth – A Report for THE CLUB OF ROME'S Project on the Predicament of Mankind*. New York: Universe Books.
Mill, John Stuart. 1848. *Principles of Political Economy with Some of Their Applications to Social Philosophy*. C.C. Little & J. Brown.
Mirowski, Philip. 1989. *More Heat than Light: Economics as Social Physics, Physics as Nature's Economics*. Cambridge: Cambridge University Press.
Mishkin, Frederic. 2004. *The Economics of Money, Banking, and Financial Markets*. Upper Saddle River: Prentice Hall.
Molina, Mario, McCarthy, James, Wall, Diana, et al. 2014. *What We Know: The Reality, Risks and Response to Climate Change*. Washington, DC: The American Association for the Advancement of Science.
Monsanto. 'Monsanto | Our Commitment to Sustainable Agriculture', available at http://www.monsanto.com/
Obama, Barack. 2013. *Remarks by the President on Climate Change*. Georgetown University: The White House.
OECD. 2002. *Sustainable Development: Indicators to Measure Decoupling of Environmental Pressure from Economic Growth*. OECD.
OECD. 2011. *Towards Green Growth*. OECD Documents on Green Growth.
Pierre-Louis, Kendra. 2012. *Green Washed: Why We Can't Buy Our Way to a Green Planet*. Brooklyn, NY: IG Publishing.
Quesnay, François. 1758. *Tableau Économique*. London: Macmillan.
Ricardo, David. 1817. *The Principles of Political Economy and Taxation*. London: Dent.
Robin, Marie-Monique. 2010. *The World According to Monsanto: Pollution, Corruption, and the Control of the World's Food Supply*. New York: New Press.

Rockström, Johan, Steffen, Will, Noone, Kevin et al. 2009. 'Planetary Boundaries: Exploring the Safe Operating Space for Humanity'. *Ecology & Society* 14(2).
Romer, Paul M. 1986. 'Increasing Returns and Long-Run Growth'. *Journal of Political Economy* 94(5): 1002–37.
Rowbotham, Michael. 1998. *The Grip of Death: A Study of Modern Money, Debt Slavery, and Destructive Economics*. Concord, MA: Jon Carpenter.
Ryan-Collins, Josh, Greenham, Tony, Werner, Richard and Jackson, Andrew. 2011. *Where Does Money Come From? A Guide to the UK Monetary and Banking System*. London: New Economics Foundation.
Sachs, Wolfgang. 1999. *Planet Dialectics: Explorations in Environment and Development*. London: Zed Books.
Sloterdijk, Peter. 2011. *Spheres. Vol. 1: Bubbles: Microspherology*. Los Angeles, CA: Semiotext(e).
Smith, Adam. 1759. *The Theory of Moral Sentiments*. New York, NY: Penguin Classics.
Smith, Adam. 1776. *The Wealth of Nations*. New York: The Modern Library.
Solow, Robert M. 1956. 'A Contribution to the Theory of Economic Growth'. *The Quarterly Journal of Economics* 70(1): 65–94.
Swan, Trevor W. 1956. 'Economic Growth and Capital Accumulation'. *Economic Record* 32(2): 334–61.
Tansley, A. G. 1935. 'The Use and Abuse of Vegetational Concepts and Terms'. *Ecology* 16(3): 284–307.
Tanuro, Daniel. 2009. *Green Capitalism: Why It Can't Work*. Halifax: Fernwood Publishing.
Turgot, Anne-Robert-Jacques. 1774. *Reflections on the Formation and Distribution of Wealth*. London: E. Spragg.
United Nations. 1987. *Report of the World Commission on Environment and Development: Our Common Future*. New York: United Nations.
United Nations. 2012. *United Nations Rio+20 The Future We Want*. Rio de Janeiro: United Nations.
United Nations Environment Programme. 2011. *Decoupling Natural Resource Use and Environmental Impacts from Economic Growth: An Excerpt of the Report*. Paris: UNEP.
Vernadsky, Wladimir I. 1944. 'Problems of Biogeochemistry II'. *Transactions of the Connecticut Academy of Arts and Sciences* 35: 483–94.
Weber, Max. 1904. ' "Objectivity" in Social Science and Social Policy'. In *The Methodology of the Social Sciences*. New York: Free Press.
Werner, Richard. 2005. *New Paradigm in Macroeconomics: Solving the Riddle of Japanese Macroeconomic Performance*. Basingstoke: Palgrave Macmillan.
Wilkinson, Richard G. and Pickett, Kate. 2010. *The Spirit Level: Why Equality Is Better for Everyone*. London and New York: Penguin Books.
Wittgenstein, Ludwig. 1918. *Tractatus Logico-Philosophicus*. Mineola, NY: Dover Publications.

World Bank. 'GDP per Capita, PPP (current International $) | Data | Table.' Available at http://data.worldbank.org/indicator/NY.GDP.PCAP.PP.CD?order=wbapi_data_value_2005+wbapi_data_value&sort=desc&page=1

Yuran, Noam. 2014. *What Money Wants: An Economy of Desire*. Stanford, CA: Stanford University Press.

Žižek, Slavoj. 1989. *The Sublime Object of Ideology*. London: Verso.

Žižek, Slavoj. 1991. *Looking Awry: An Introduction to Jacques Lacan through Popular Culture*. Cambridge, MA: The MIT Press.

Žižek, Slavoj. 1997. *The Plague of Fantasies*. London: Verso.

Žižek, Slavoj. 1999. *The Ticklish Subject: The Absent Centre of Political Ontology*. London: Verso.

Žižek, Slavoj. 2000. *Contingency, Hegemony, Universality: Contemporary Dialogues on the Left*. London: Verso.

Žižek, Slavoj. 2000. *The Fragile Absolute Or, Why Is the Christian Legacy Worth Fighting For?* London: Verso.

Žižek, Slavoj. 2003. *The Puppet and the Dwarf: The Perverse Core of Christianity*. Cambridge MA: The MIT Press.

Žižek, Slavoj. 2006. *How to Read Lacan*. London: W. W. Norton & Company.

Žižek, Slavoj. 2006. *The Parallax View*. Cambridge, MA: MIT Press.

Žižek, Slavoj. 2008. 'Nature and Its Discontents'. *SubStance* 37(3): 37–72.

Žižek, Slavoj. 2009. *In Defense of Lost Causes*. London: Verso.

Žižek, Slavoj. 2009. *Violence: Six Sideways Reflections*. London: Profile Books.

Žižek, Slavoj. 2011. *Year of Distraction*, Lecture, available at https://www.youtube.com/watch?t=17&v=ChWXYNxUFdc

Žižek, Slavoj. 2012. *Less than Nothing: Hegel and the Shadow of Dialectical Materialism*. London and New York: Verso.

Index

abundance 77, 103, 105
Adam 79–83, 163–4
   Adam and Eve 24–5, 174–5, 181, 183, 187
Adorno, Theodor W. 6, 9, 44
American Association for the Advancement of Science (AAAS) 148–50
animal(s) 7, 14–33, 50, 57–9, 71, 122, 155, 206
   Man as polluting animal 19, 24–6, 32, 38, 49, 51
Apollo 17 33, 36–8, 43, 228
apple 24, 80, 83–4, 104, 164
Apple 83–4
autark 169–70

Bachelier, Louis 32
bank 132, 138, 158, 197–226
   Bank of England 197
   bank as financial intermediary 199
   Bank of International Settlements (BIS) 138
   central and commercial 197–201, 218–19
   World Bank 128, 130–1, 138
banking *see* bank

Barro, Robert J. 111, 135, 138
barter 82, 130, 164–5, 178–81, 190, 201, 203
Baudrillard, Jean 86–9, 104, 126–7, 220
Berry, Wendell 184–6
big Other 40, 42, 46–7, 63, 90, 123, 185, 191, 201
biogeochemistry 29–32
biosphere 30–3, 42, 56, 106–7, 159
blé 66–7, 73, 151, 158, 187, 196
Boltanski, Luc 181
Bretton Woods 197–8
Brundtland, Gro H. *see* Brundtland Report
Brundtland Report 33–45, 151–4, 157–60, 168
Buffon, Comte de 20, 23

capital 17, 22, 32, 46, 71–2, 76–7, 93–102, 106–14, 117–18, 120–1, 125–9, 132–8, 178–9
   accumulation 93, 96–8
   capital and labour 22, 106, 109–14, 117, 121, 125–6, 129, 134–5, 138, 194, 230

capital (cont.)
  depreciation 96
  formation 72, 76
  investment 72, 109
  money as capital 192
  natural 106, 134, 179, 230
  requirements 199, 202
capitalism 1, 3–5, 23, 39–41, 46, 81, 92, 128, 132, 144–6, 150, 171, 192–4, 223, 226, 231, 233
  beyond growth 61
  consumer 41, 46, 83, 119, 193
  critique 144–6, 181–8, 231–2
  global 46, 104, 229
  green 1, 4, 41, 144–5, 151, 160, 181–2, 233
  growth 10, 41, 187, 190, 193–4
  industrial 21, 74, 119
  neo-liberal 183
  spirit of capitalism 144, 181
catastrophe 3–4, 13–14, 24, 31, 38–43, 48–9, 118, 146, 149, 231
Chaplin, Charlie 132
Chiapello, Eve 181
Christian 24, 50–1
Christianity *see* Christian
Clements, Frederic 29
climate change 1, 3–4, 13–14, 23, 33, 35, 43, 45–8, 57, 89, 143–50, 173, 196, 228, 231
CO_2 14, 27, 56–7, 89, 106
commodity 59, 61, 73, 85, 88–9, 94–6, 103, 121, 127, 137, 165–8, 178, 189
concrete universality 131
conspiracy 227
constructivism 9
consumption 1, 3–4, 7, 14–15, 25, 39–41, 50, 61–2, 66, 73, 82–3, 99–104, 146, 154, 158, 162–4, 169–70, 186, 190, 195–6, 221, 229–33
  GDP 128–9, 134, 162–3
  objective and subjective 123–7

production function 95–7, 101, 108, 114–22, 169
consumptivity 121, 125–7
corporation 152, 154, 219
crisis 10, 14, 188, 190, 197
  ecological 1–4, 38, 92, 96, 144–5, 151, 160, 227, 233
  economic/financial 3, 10, 13, 46, 143–6, 199, 233
cynicism 36, 132

Daly, Herman 99
Dandalos, Nick 'the Greek' 194
death 16, 20, 26, 38–40, 58, 179, 225–6
debt 3, 57, 136–7, 194, 197–9, 202–28, 231
decoupling 16, 145, 194–200, 222–3
demand 14, 118–27, 136, 155–7, 163, 189–90, 194–5, 199, 201, 205, 217, 222, 231
  supply and demand 22, 83–90, 167
Derham, William 15–16, 21
desire 73, 80–3, 91–2, 113–24, 147, 150, 155–8, 165–8, 184–5, 191–4, 197, 201–3, 209, 210, 212, 223–5; *see also* gross domestic desire (GDD)
drive 120, 145, 192–4, 203, 223, 225–6

Earth 13, 30, 33–7, 42–3, 88–9
eco 5–10, 15, 17–19, 24–7, 37, 42, 44–5, 48, 58, 62, 67, 75, 82, 86–108, 146–51, 176, 178, 181, 190, 196, 227, 229
eco-analysis 3–9, 15, 18, 42, 44, 55–9, 61–2, 67, 75, 82–3, 91, 95, 114, 116, 125, 128, 145–6, 150–2, 163, 165, 174, 178, 187, 194, 212, 227, 229, 232–3

Index

eco-naming 67, 75, 87, 95, 98, 102
economics
  classical 22, 74–6, 86, 115
  ecological 14, 97, 99, 106–7, 146, 183, 187, 196, 230
  neo-classical 32, 49, 63, 85–97, 101–2, 106–7, 115–16, 120, 126, 128, 137, 165, 176, 180, 187, 190, 229, 231
  physiocrat 62, 66, 69–70, 74, 76, 86, 98, 151, 155, 187, 196
ecosystem 28, 33, 43, 56–7, 99, 105–6, 144–5, 148
education 110, 130, 134, 184–6
efficient-market hypothesis (EMH) 32–3
Eisenstein, Charles 205
emergentism 31
Engels, Friedrich 1–3
enjoyment 80, 123–7, 164
equilibrium 9, 96–7, 110, 137
exchange
  equivalent exchange 75, 85–9, 103–7, 137, 156, 167, 175–81, 187, 220
  impossible exchange 86, 9, 102, 104–5, 108, 156, 167–8, 176–7, 181, 187, 220
  medium of exchange 164–7, 179, 190
extinction 14, 20–1, 23–5, 35

Facebook 120
Fama, Eugene 32
farming 56–7, 63, 65–6, 72–8, 102, 159, 179–80, 203–4, 206, 212–14
feminism 182, 187–9
food vs fuel 126
Forbes, Stephen A. 28–9
Ford, Henry 118–19
Foucault, Michel 8, 19
fracking 27, 233
Fraser, Nancy 182, 188

Gaia 13, 33
gambling 193–4
genetically modified organism (GMO) 152, 159–60
Georgescu-Roegen, Nicholas 196
gift 106, 130, 134, 165, 175–81, 187, 91, 205, 215
  gift of nature 70–2, 75–9, 102, 104, 108, 187
global warming *see* climate change
God 15, 17, 22–4, 36–7, 51, 83, 175
Golden Rule of Capital Accumulation, The 97
Gore, Al 41, 233
Graeber, David 135
greed 223, 226–7
gross domestic desire (GDD) 114, 116, 120–1, 127
gross domestic oroduct (GDP) 57, 60–3, 89, 109, 118, 120–2, 126, 128–39, 162, 170, 176, 181, 183–4, 187, 209
growth
  green 143–6, 150–1, 162–3, 169–74, 195, 232–3
  horizontal and vertical 130
  imperative 41, 46, 58, 69, 81, 146, 151–4, 160–3, 186, 197, 205, 217, 223, 231
  limits 39, 61, 92, 101, 109, 145, 148, 229
  perpetual 57, 115, 145–6, 155, 173–4, 191, 196–7, 205, 221, 225
guilt 50–1

Haeckel, Ernst 21, 31
Hegel, Georg W. F. 131
Heidegger, Martin 42, 59–61, 95
Herodotos 15
homo economicus 169–70
household 5, 65, 72, 76, 108, 170–2, 182–9
housekeeping 171, 183–5

Index

ideology 10, 14–15, 26–7, 37, 40, 42, 49–51, 56–7, 86, 128, 133, 138, 144–5, 179–80, 186, 191, 193, 195, 229
imaginary 4, 6, 25–7, 37, 40, 62–3, 86, 90–1, 95–6, 102–3, 117, 133, 188, 208, 210, 217, 221
interest (on debt) 57, 100, 136–7, 197, 202–6, 209–22
International Monetary Fund (IMF) 55–60, 138
interpassivity 122–3
investment 32, 35, 72, 76, 94, 96–8, 107–10, 113, 121, 128, 138, 144, 162, 166, 186, 212, 215, 218, 222
invisible hand 16–18, 22, 79

Jackson, Michael 44
Jiabao, Wen 47
job(s) 106, 173–94, 203, 215
  green 172–4, 190

Keynes, John M. 115, 201–2
Keyser Soeze 158–60
Kuznets, Simon 132
Kyoto Protocol 51

labour
  capital and labour 22, 106, 109–17, 121, 125–6, 129, 134–5, 138, 230
  division of 16, 38, 65, 79–83, 136, 146, 163–83, 187, 193, 201
  maternal 175–6, 183, 187–8
Lacan, Jacques 6–7, 9, 18, 46, 68, 75, 97, 134, 150, 155, 157, 166, 168, 178, 187, 201, 209, 212
land 62–3, 66, 69–71, 76–8, 97, 100–3, 107, 153–4, 159, 175, 217
Leclerc, Georges-Louis 20
Leibniz, Gottfried W. 23

Lenin, Vladimir 1–10, 58, 144, 227, 230, 233
Lietaer, Bernard 203–8
lifestyle 14, 41, 50, 124, 146–7, 161
Linneaus, Carl 15–21
logos 5, 8, 55
Lovelock, James 33

Man 21–7, 30–1, 35, 37, 49, 60, 77–9, 82, 154, 229
Mankiw, Gregory N. 93, 129, 130–2, 139
market
  efficient 32–3, 63, 86, 89–90
  financial 32, 222
  free 49, 69, 74, 85, 97
  labour 183–5, 193
  marketplace 38–40
  rational 74
Marsh, George P. 21, 23
Marx Brothers 112, 207
Marx, Karl 1–3, 21–2, 92, 134, 137, 224, 228, 231
materialism 166, 228, 230
Maugham, Somerset 38–9
Mauss, Marcel 188
metaphysics 29, 95, 120, 149, 229
Mill, John S. 72, 178
money
  credit 198–203
  desire 166–72, 192–4
  fiat 197–200, 219
  veil 178–9
Monsanto 152–4, 158–61
mOther 155–7
Mother Nature 51, 70–1, 76, 79, 102, 187

natural resources 14, 27, 35, 62, 96–7, 102–8, 132, 136, 139, 143, 154, 159, 176, 185
natural selection 20–3
need 83–4, 119, 155–7, 168, 193
nomos 5, 8, 55, 85, 87, 171, 183
noosphere 30–2

Obama, Barack 46–7, 56, 144, 172–3, 231
object
  desire 81, 114, 185, 193, 201, 212
  objet petit a 112–13, 129, 135, 166–7, 208
  split 19
  sublime 131, 193
Occupy Wall Street 223
oikos 5, 8, 44, 87, 171, 183
ontology 23, 37, 50, 228–9
Organisation for Economic Co-operation and Development (OECD) 144–5, 172, 195

paradox 13, 24, 26, 28, 31, 131, 149, 163, 166, 183, 190
parallax 5–9, 17–18, 24–5, 43, 68, 166, 168
Phelps, Edmund S. 96–7, 139
Phillips, John 29
point de capiton 82, 158, 168
pollution 3, 13, 23–4, 31, 35, 39, 58, 107
production function 63, 93–102, 106–17, 125–7, 136–8, 169–70, 176
property 71, 103–4
Protestantism 15, 48, 51
psychoanalysis 5, 44, 80, 91, 95, 116, 146–7, 150, 227

Quesnay, François 62–71

racism 158
Rasmussen, Lars L. 47–8
real 6–10, 18–19, 25–7
  eco 8–9, 26, 37, 75–9, 82, 88, 92, 98–102, 107, 176–80, 196
  GDP 128, 133–4
  reality 8, 26, 43, 56, 58
  value 67–73, 90, 92, 134, 137
rent 63, 76–8, 214–16
revolution 14, 35, 37, 40, 88, 93, 171, 231
Ricardo, David 72, 62, 73–8

Ritchie, Lionel 44
Romer, Paul M. 63, 110–11, 115–20, 127, 139
Rumsfeld, Donald 117–18

Sala-i-Martin, Xavier I. 111, 135, 138
scarcity 97, 99, 103, 105, 143, 196, 204, 223
Schumacher, Ernst F. 14
Seinsvergessenheit 95
sex 169–70, 180, 182, 189, 195
Sloterdijk, Peter 149
Smith, Adam 16–17, 62, 72–4, 163–4, 175, 181–3
Solow, Robert 63, 93–8, 109–16, 120
speculation 32
state 49, 63, 69, 134, 185–7, 190, 197–8, 200–1
steady state 96–8, 109–10, 115
subjectivity 6, 18–19, 38, 40, 49, 80, 82, 158, 160, 168, 193, 207, 225
  split subject 18, 25, 40, 95, 168–9, 207–8, 225
  subject of ecology 25
  subject of labour 75–6, 113, 129
sun 37, 43, 57, 103
superorganism 29
supply *see* demand
sustainability 10, 34, 36, 40, 67, 105, 143–6, 151, 154, 158–9, 181
Swan, Trevor W. 63, 93–8, 109–16, 120
symbolic
  castration 81, 120, 156–7, 163–9, 185, 189, 193, 207, 212
  order 6–10, 18–19, 25–6, 40, 42, 48–9, 61, 72–5, 80–3, 98, 125, 156–7, 164, 177, 189–90, 200, 207, 210, 212
  order of economy 95, 104–8, 120, 156, 167, 170, 187
symptom 6, 68–9, 186, 190

Tansley, Arthur G. 28–9
Taylor, Mark P. 129–30
theology 15–16, 21–2, 79
trauma 6, 25–7, 37, 43–5, 49,
    68–9, 86–93, 133, 168, 181,
    200, 212
traumatic *see* trauma
Turgot, Anne Robert
    Jacques 62–71

United Nations (UN) 34, 37, 40
United Nations Climate Change
    Conferences (COP) 45–8

value
  exchange 134, 137, 167
  and price 10, 62, 67–9, 86,
    89–90, 168, 229
  surplus 65–6, 70, 74, 79, 113
  use 134, 137, 166, 217

Vernadsky, Vladimir 29–32
violence 104, 188–9

waste 27, 31, 41, 106–8, 170–2,
    176, 185–6, 195
wealth 14, 17, 60, 63, 65, 73–4,
    92, 115, 129–32, 139, 158,
    186, 190, 210, 220, 224
Wittgenstein, Ludwig 8, 137
World Trade Organization
    (WTO) 138

Yuran, Noam 224, 232